ELEMENTAL INDIA

ELEMENTAL INDIA

THE NATURAL WORLD AT A TIME OF CRISIS AND OPPORTUNITY

Meera Subramanian

HARPER
LITMUS

First published in hardback in India in 2015 by
HarperCollins *Publishers* India

Building 10, Tower A, 4th Floor, DLF Cyber City, Phase II,
Gurugram Haryana – 122002, India
www.harpercollins.co.in

P-ISBN: 978-93-5136-528-0
E-ISBN: 978-93-5136-529-7

2 4 6 8 10 9 7 5 3 1

Typeset in 11/14 Nebraska Regular at
SURYA

Printed and bound at
MicroPrints India, New Delhi

Dedicated to my parents, Mani and Ruth Subramanian

Life,
When elements become ordered, that's all.
Death,
*But a moment when into chaos they fall.**

زندگی کیا ہے عناصر میں ظہور ترتیب
موت کیا ہے انہیں اجزا کا پریشاں ہونا

—Brij Narain Chakbast,
Urdu poet

*Raza Mir, ed. and trans., *The Taste of Words: An Introduction to Urdu Poetry* (Gurgaon: Penguin Group, 2014), 108, http://www.outlookindia.com/printarticle.aspx?292377.

CONTENTS

INTRODUCTION

When he was little more than a boy in Madras, my father learned of the family astrologer's prophesy that he would grow up and one day cross the Kala Pani, the demon-filled Black Waters that separate India from the rest of the world. Hindus, especially high-caste Brahmins, were forbidden to traverse such divides, yet the prediction came true. When he was twenty-five, my father became the first in his family to leave South Asia. He travelled by ship for three weeks and a day before arriving in New York Harbor. It was 1959, and he planned to stay for eighteen months in pursuit of a master's degree in engineering. Five years later, he had acquired a PhD, to his family's delight, and an American wife, to their distress. He was part of that first trickle of Indian immigrants that would grow into a flood, and his union with my mother was representative of what would become, a half-century later, an intimate link between cultures. Then, Asian-Indian wasn't even a category on the US census. By 2014, yoga had become a common American pastime and a visit from the Indian prime minister could pack Madison Square Garden in the heart of Manhattan.

After my brother and I were born, my father took all of us to Madras to meet his family. Though I was too young to remember that first visit, the arms of aunties and grandparents who cradled me surely imprinted, and I still remember exquisite details from later trips. My father's vast clan in India—to whom I was closer than to the small scattering of American relatives of European descent on my mother's side—sowed the seeds of my interest in India's natural history and her people. Their faraway influence broke through the thick crust of my otherwise insular suburban upbringing in New Jersey (malls! rock concerts!), where my parents had settled to raise my brother and me.

On each trip to India, I gathered impressions. There were the seeds of the mehendi plant whose leaves an aunt ground into a paste to decorate my hands for a cousin's wedding when I was ten. There were seeds that grew into the limes I plucked from an uncle's front yard in Alwarpet to make fresh juice at his house when I was thirteen. And there were those from the neem tree whose upper branches my grandmother reached for from her balcony and used to craft me a toothbrush when I was nineteen. Maybe it was these seeds that planted in me a desire to leave the race of the upwardly mobile north-east and landed me at the end of a dirt road in Oregon by the time I was twenty-six.

Aprovecho Research Center, the environmental non-profit organization where I lived and worked for nearly a decade, was composed of an eclectic team of idealists, hungry for knowledge. Much of the 40-acre land trust was a forest where we collected the wood to build and warm our homes and to cook the food we grew organically. We harvested electricity from the sun with a solar array and built a small dam to trap water from a spring. Both a school and research centre, Aprovecho drew people from around the world to teach and learn the skills of sustainability.

But just as I was turning my back on consumerism, living deliberately in developing-world style on the edge of my developed nation, Indians were getting their first real taste of material wealth and wanted more.

In the summer of 1991, India launched initiatives that opened her economy to the world. The result was a rapid blooming, as awkward and astounding as human adolescence. After nearly half a century as a borderline socialist nation—with one inept state airline and two types of cars to which you could affix a 'Be Indian Buy Indian' bumper sticker—this still young country had vaulted into the global economic fray. The once wealthy 'Bird of Gold', as it was called before centuries of colonial rule plucked her bare, was again poised to soar.[1] Analysts predicted that the world's largest democracy would soon outpace China economically and become the fifth largest consumer market by 2025—with a middle class that would exceed the entire population of the United States.[2] As the United States and the European Union struggled to regain their economic footing after the 2008 economic crisis, the BRICS—Brazil, Russia, India, China and South Africa—continued to rise, with India leading the way. That 'Western power will remain intact', wrote Patrick French in *India: A Portrait,* in 2011, 'is an outdated and fantastic view'.[3]

But as India travels on this path of progress, masses of Indian citizens are being left behind, and the lands and waters that have sustained her people for millennia are being compromised. Since achieving Independence in 1947, India's population has tripled. While a rapidly growing middle class has gained unprecedented material comforts, only a tiny fraction of Indians can afford to shop at the glitzy air-conditioned malls springing up across the country. Today the vast majority of Indians still struggle to meet fundamental daily needs. Six out of ten citizens lack access to clean water, a third live beyond the glow of the electric grid, and less than half have access to a toilet.[4] More than two-thirds of households cook over an open fire,[5] and the food they prepare there is inadequate: nearly half of all children under five are stunted, and the rate of malnutrition currently exceeds that of sub-Saharan Africa.[6]

Sustainable development is said to be the solution, but what does this buzz term mean? How can it help farmers tilling their soil in Punjab or mothers preparing dinner in Maharashtra? How can it affect biodiversity in cities, in wild reserves and in rural areas?

A catastrophic collapse of the South Asian vulture population led me back to India as a reporter in 2009. The story revealed complex relationships among India's exponentially growing population, the finite resources that sustained them, and religious and cultural circumstances wholly unique to India but echoed around the world. A couple of years later, over a Punjabi meal of dal, sabzi, rice and nan in Boston with friends Indian and American, the idea of a book devoted to stories of India's land and the people working to sustain it emerged. I would tell five stories, one for each of the five elements: earth, water, fire, air and ether.

The elements, what they are and what they signify, are as diverse as the human cultures that have wrestled to identify them throughout history. Ancient Greek philosophers ruminated endlessly on what constituted the foundation of the physical world. Plato used the word 'element', and Aristotle classified earth, water, fire and air as either wet or dry, hot or cold, adding ether to the mix.

In Islam the elements, anāṣir, are seen as prime matter, material indivisible, with earth farthest from heaven and fire closest.[7] In ancient Tibetan philosophy, the five elements equate with the five senses, the primary pranas—vital energies—of existence.[8] In Chinese belief systems, the five elements of Wu Xing are seen as ever-changing forces or phases, rather than immutable building blocks of matter.[9]

The elements also appear in the Upanishads as the pancha mahabhuta, where each is an integral part of prakṛti, nature. Each finds its expression within the prayer ceremonies I witnessed in India as a child, when the smell of flowers and fire would intermingle with the hypnotic sound of chants as water was poured by priests rooted on the ground. The *Taittiriya Upanishad* places the elements in relation to the divine Brahman with these words, like the series of 'begats' in the Jewish and Christian bibles: 'From Brahman arises ākāśa (ether), from ākāśa arises vayu (air), from vayu arises tejas (agni / fire), from tejas arises ap (water), and from ap arises prithvi (earth).'[10]

In all their variations, these elements form the world we all inhabit. They play off each other in an eternal dance of equilibrium. But human impacts have increasingly upended this natural balance. Today, equipoise among the elements is disintegrating, and if these ecological challenges are not tackled creatively, India's economic growth will be crippled, and her people and wildlife will continue to suffer. India has become the staging ground for an experiment in human survival. Can 1.2 billion people, one of every five on earth, inhabiting a mere 2 per cent of the world's land mass, learn to live sustainably?

It is only a matter of time before even the most comfortable of countries will face similar circumstances. India's current ecological problems reflect what other nations will be confronting increasingly in the years to come, if they are not facing it already. In the United States, storms such as Hurricane Sandy on the east coast, withering droughts in California and forest fires throughout the west are growing ever more frequent and devastating. In the Middle East, water scarcity exacerbates conflicts between Syria, Iraq and Lebanon.[11] Sub-Saharan Africa's population is poised to double by 2050.[12] India's mistakes and successes could provide a map forward for the rest of the world.

Now is the moment when India can choose what direction she will take. What inspiration will guide this South Asian nation, as she develops her people and her resources? I criss-crossed India in search of answers.

The first element, earth, led me to Karnataka, Punjab and

elsewhere to explore where the human and ecological tolls of the Green Revolution are visible after forty years of intensive chemical agriculture. Water took me to the semi-arid landscape of Rajasthan, where dried-up wells are being reinvigorated by villagers who lift their shovels and pickaxes to transform the land. I found the element of fire burning in cooking stoves fuelled by wood or dung and lit every day in most Indian households. The vultures that no longer flew through the air over India led me to a vulture-breeding centre in Haryana, a carcass dump overrun by dogs in Bikaner and the sacred grounds in Mumbai where Parsis once laid out their dead on Towers of Silence for vultures to consume. Ākāśa, space or ether, called me to investigate population growth in Bihar, the state with one of the nation's highest fertility rates and some of its youngest brides.

Yet none of these stories could be confined by the terms I designed for them. To consider farming is to explore water as much as earth. The smoke from wood-fuelled fires blackens the air and the lungs of women and girls. Each element led to the next, all inextricably linked.

The challenges facing India provide myriad reasons for despair, and some of the stories I found are more cautionary than celebratory. But in unexpected places I witnessed seemingly powerless people revitalize their environments and their lives. Across the subcontinent I met people who believe a better India is possible: an engineer-turned-farmer brings organic food to Indian plates; the award-winning Rainman of Rajasthan and villagers help a river run again; well-intentioned designers of cooking stove relentlessly pursue a smoke-free fire; a passionate pair of biologists coax vultures to reproduce; a young woman teaches girls about their bodies and rights as stories of sexual violence increasingly emerge across India.

Some experiences only revealed their significance to me months or years later. In the gloaming of an autumn night in

Rajasthan, on the way to an interview, I stumbled upon a village ceremony. Forty men sat on a large cloth that lay on the ground between two long rows of houses. The grey relief of the Aravalli hills towered behind them as the last of the day's light singed the sky. A man in his thirties sat before the group as an older man behind him wrapped rounds and rounds of white fabric upon his head. The mass of material fast became an enormous turban of nearly farcical proportions, more than two feet across and a foot and half tall.

The men sitting around me passed around hand-rolled beedis and smoked them. One by one, they handed a few rupees and their own contribution of white cloth to a man who would note it in a record book, and then pass it to the older man, who incorporated the material into the great turban by linking one piece of material to the next, ever winding.

Later, I learned that the younger man's father had died. The community members gathered were passing on the respect they once held for his father on to him. The weight of the cloth on the man's head was immense. He could barely stand when the ritual was over but he was lifted upright with the help of the men of the village. The women who had been standing on the periphery drew nearer, singing. Together, they all led him away downhill and faded like a mirage into the night.

The image stayed with me after I returned to the United States. It haunted me as friends' fathers died, and family and friends died too. I had never witnessed such a profound expression of the emotional burden one carries upon the death of a parent, but that cumbersome cloth—so precarious, so weighty, almost comic—seemed to represent more.

As I investigated stories of economic progress and of blackouts and gang rapes, the more I experienced what I imagined that man had felt under his heavy burden. I carried the stories of the people I met, and felt the responsibility to share them. This Earth also suffers, carrying the load of a precocious species that has proven the ability to despoil every recess of our land, air and water but also holds the power to create other ways to sustain ourselves, should we choose. Each of us helps to shoulder this weight in some way. We are all accountable to the earth, humanity and our shared future. India's burden is particularly heavy. This book is her story.

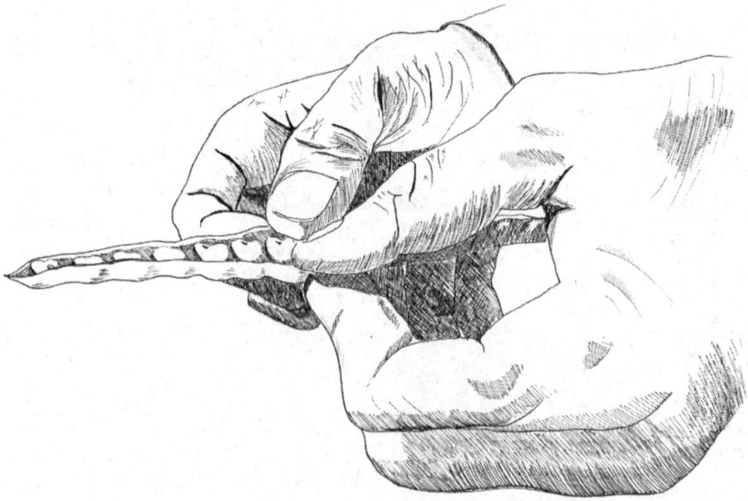

1

PRITHVI

EARTH

There are no miracles in agricultural production.

—Norman Borlaug,
father of the Green Revolution[1]

SILENT SPRING

'It looked like sugar.' White. Crystallized. Tantalizing. That was Swaran Singh's memory of the first time he saw pure nitrogen fertilizer. 'I remember the gram sewak,' he said, referring to the village officer, 'coming to the house with a cart full of urea, offering it for free.'[2] It was like crack for crops.

'Don't take anything they give you for free,' Swaran Singh's sceptical grandfather warned. 'It's like the tea that the British gave us, and now it's like a drug.' But it was the 1960s, and Swaran and his brothers were young men, excited about the new chemistry, modern farming, the potential of a technological future and the hope of release from the ceaseless cycles of agrarian life.

Swaran was beanpole thin when I met him, dressed in a white cotton kurta and an emerald green turban. His legs were so slender that when crossed, they looped around twice, one foot hooking under the opposite ankle. He had unruly grey-tipped eyebrows, a white beard and deeply set eyes with irises ringed with the blue of early cataracts. We sat together in the courtyard of his house in Karamgarh, a village his forefathers had established 160 years earlier, south-east of the city of Bathinda, a few hours from the Pakistan border. A neem tree that Swaran's father planted at the family homestead in 1951 to mark Swaran's birth shaded us from the sun, and a creamy-white cow that no longer gave milk but remained a precious commodity for its manure stood a few feet away. Swaran's family had farmed for six generations without chemicals, until the carts of urea, a nitrogen fertilizer, arrived on their doorstep.

It was the dawn of the Green Revolution, a post–World War II initiative that introduced widespread use of high-yield, water-intensive and fertilizer- and pesticide-dependent crops to farmers across Mexico, the Philippines, India and the world. Factory-made nitrogen, first developed in 1909 by a German chemist named Fritz Haber, captured the elusive element that is essential for plant growth from the air.[3] This synthetic nitrogen proved essential for bomb making during modern wars and for growing food in modern times. It completely transformed agriculture in the twentieth century. So much so that today, almost half of the nitrogen found in the muscle and organ tissue of our bodies originated in a fertilizer factory.[4]

In the beginning, the urea and the aerial sprayings of pesticides cost the farmers nothing, and the hybrid wheat seeds that arrived from the West were subsidized.[5] The government drilled bore wells across the land of Punjab, and water gushed out of pumps run with free government electricity.[6] Farmers had nothing to lose.

Punjab is named for the five rivers that run through her (punj means five and ab is water in Persian): the Jhelum, the Chenab, the Ravi, the Beas and the Sutlej. All are tributaries to the mighty Indus river, but collectively they were still not adequate to provide for the needs of Punjab, the breadbasket of India, during the Green Revolution. The boundless irrigation combined with those first applications of synthetic powders worked magic on the fields. Instead of leaving fields fallow, patiently looking skyward for the coming of monsoon rains, farmers planted successive crops, rounding out the days between the wet kharif and dry rabi seasons. Those were heady days, times of growth and promise of more after perennial seasons of famine and near-famine that had visited and revisited South Asia for centuries.

'When grandfather found out what we'd done,' Swaran Singh said, 'he told us we'd regret it.'

International agribusiness companies targeted this small north-western state forty years ago and today Punjab produces nearly a fifth of the nation's wheat[7] and 42 per cent of its rice,[8] though it inhabits a mere 1.5 per cent of India's land mass. It also accounts for 17 per cent of the country's pesticide use,[9] and now the landscape is as silent as Rachel Carson's unnamed town in *Silent Spring*, eerily bereft of the mewing calls of peacocks or the songbirds that were once abundant. In the fields, women and children pluck cotton with nimble fingers for about a hundred rupees per day, while men walk barefoot through the rows with pesticide sprayers lashed to their backs. Where productivity soared for several generations with the thick application of pesticides and fertilizers, yields are now flat.[10]

Since the Green Revolution's arrival in the 1960s, four billion people have been added to the world's population, seven hundred million of them in India—a country that has experienced ninety famines in 2,500 years of recorded history.[11] The last occurred in 1943 in Bengal, when several million people perished. It led Jawaharlal Nehru, India's first prime minister, to declare soon after Independence in 1947, 'Everything else can wait, but not agriculture.'[12]

Another spectre spurring the Green Revolution was the memory of an India dependent on imported food.[13] In 1967, millions of Indians relied on foreign aid to eat, but by 1991 India was more than self-sufficient in food production, having doubled its agricultural returns during the 1970s and 1980s[14] because of Green Revolution methods. Yields of rice and wheat multiplied.[15] But imports had not ceased. They had merely shifted form. Instead of foodstuffs, India brought in foreign fuel, farm equipment, fertilizers, chemicals and seeds. The research and development behind each new advance came from outside of India's borders. Is India's much lauded achievement of food self-sufficiency a deception, her reliance on food imports swapped for an addiction to fertilizers and patented seeds that must be purchased from abroad season after season?

In terms of feeding people, the Green Revolution did work. Its methods quickly became seen as conventional, though there was little that was traditional about them. They provided a level of food security and food sovereignty for the young nation. But at what cost? India holds a fifth of humanity in her embrace. Is it madness to explore alternatives to the package the Green Revolution presented?

The question is urgent. By 2030, the Food and Agriculture Organization (FAO) of the United Nations predicts the world will need to generate 35 per cent more calories than we do today just to feed ourselves.[16] Add to that the 2014 prediction by the Intergovernmental Panel on Climate Change (IPCC) that crop yields are expected to decline over the next century, some dramatically,[17] as temperatures rise. Relatedly, India's worsening air pollution—skies clouded with black carbon, which affects both precipitation and radiation, and ozone, which is directly toxic to plants—has been found to cut wheat and rice yields by half in some places.[18] A changing climate will bring with it both drought and deluge, or dangerous swings between the two, along with even higher temperatures.[19]

The farmers in Punjab are standing at a crossroads. Many maintain that they stepped into a bright technological future, and they refuse to look back. They say that stories about epidemic rates of cancer, polluted waters and decreasing crop yields are overblown bits of collateral damage in the war against hunger, exaggerated by alarmist media and rabble-rousing activists. But a growing number are turning to organic agriculture, from lifelong farmers who've never left their fields to educated idealists with advanced degrees and romantic agricultural ambitions who have arrived on the land from cities. They are shunning synthetic fertilizers created in factories and seeds spliced with genes in labs, and forgoing chemical controls to keep weeds and pests at bay. Instead, they are choosing to improve the overall fertility of their farms' ecosystems, focusing on good tilth—the fundamental health of the soil—and using biological controls such as natural beneficial predators to fight the pests that threaten their crops.

Could India begin to emerge from its silent spring?

AS PURE AS MOTHER'S MILK

On 17 July 2013, schoolchildren in the village of Dharmasati Gandawa in Bihar leaned over their stainless steel plates to scoop up the food served by India's school lunch programme. India's Midday Meal Scheme feeds 120 million children each day.[20] For many of India's poor, it's the best—and sometimes only—meal they get all day. But to the children in Bihar, the food tasted bad. They complained. One cook alerted the headmistress that the food tasted funny and was told that everything was fine, and so the children were instructed to eat. They obeyed.

Within twenty-four hours, twenty-three of the children were dead and dozens[21] more sickened. Investigations revealed that the oil used to prepare their potato and rice curry had been kept in a container that once held the agricultural pesticide monocrotophos.

This devastating event in Bihar revealed a nationwide problem that stems from the wide use of biocides in myriad forms, in cities and villages, in homes and fields. India today resembles the United States of the 1950s, an era when chemicals were widely under-regulated, and safety protocols from protective gear to safe handling were non-existent or minimal and easily disregarded. The poor printing quality of India's pesticide labels often renders specific application and proper use instructions illegible, and with only 63 per cent of adults literate, even the clearest directions are often useless. Since some chemicals did help—to reduce pests at home or in a field—it was easy to believe that using more chemicals would help more. Farmers in Guntur district in Andhra Pradesh and Warangal district in Telangana sprayed cotton up to thirty times in a season when optimum recommendations suggested half that.[22] If there was native South Asian knowledge about how to grow food, it was eagerly jettisoned during the Green Revolution. The memories of famines were too recent, the desire to move forward too strong.

But nearly eighty thousand Indians are now dying annually from the widespread use of pesticide chemicals, many of them developed in the West.[23] From 2004 to 2008, one hospital in Bathinda, Punjab, recorded sixty-one deaths from accidental inhalation of pesticides.[24] But too often pesticide poisoning is a deliberate invitation to death—suicidal farmers ingesting the same chemicals they use on their crops in order to end their lives.[25] Suicide can be a way to escape the crippling debt acquired in the new era of agribusiness, and these deaths have reached epidemic levels. By some estimates, more than a quarter of a million farmers have committed suicide in India[26] since 1995.

Other countries have banned the organophosphate monocrotophos because it has 'high acute toxicity', according to the World Health Organization (WHO), which pressured India to bar the use of the pesticide in 2009.[27] India ignored the warning, and the chemical, still legal, proliferates in India.

In 2011, India's agriculture minister Sharad Pawar acknowledged that sixty-seven pesticides prohibited in other parts of the world were widely used in India.[28] Even after a global ban on the manufacture and use of the highly toxic pesticide endosulfan was negotiated under the Stockholm Convention in 2011, India remains its largest producer and consumer. In 2014, the government grudgingly agreed to phase out use by 2017.

These life-altering chemical combinations that humans have introduced to the earth's ecosystem in the past century are everywhere. Studies have detected known carcinogens such as heptachlor and ethion in the blood of Indian citizens.[29] The insecticides endosulfan and malathion have been found, in levels far exceeding WHO standards, in the breast milk of nursing mothers.[30] Other pesticides, polychlorinated biphenyls (PCBs), flame retardants (PBDEs), dioxins and other synthetic chemicals have also been found in breast milk. Many act as endocrine disruptors, which mimic estrogens in the body and initiate a harmful cascade of biochemical changes to the endocrine system, influencing fertility and causing cancer.[31] Some of these modern-day chemicals pass from the mother to the fetus in the uterus, setting the stage for later-life cancer of the child, or even the child's children another generation down.[32]

The chemicals used on crops permeate grains, cotton and vegetables harvested from the fields. Of 345 vegetable samples collected at markets in Pune, Maharashtra, ninety-six had pesticide residues, including the banned pesticides chlordane, carbofuron and captafol.[33] Unbeknownst to the women filling up their grocery bags, nearly a quarter of vegetables in Kerala markets showed organochlorine pesticides—carrots with endosulfan undertones, curry leaves with a sprinkle of aldrin, long beans with a hint of DDT.[34] Half of the contaminated samples were over the permissible limits, rendering them unfit for human consumption.[35] In Bangladesh, the Institute of Public Health tested thousands of food items, from spices to sweetmeats. In some, adulteration rates were as high as 99 per cent.[36]

As long as the host of -cides—herbicides, fungicides, pesticides—produced to control weeds, fungi and pests in the field, and roaches, rats and mosquitoes in the city are cheap and effective, they will remain legal. They have been tested and they are safe, insist the global corporations that dominate the $30 billion market, where sales have increased thirtyfold since the 1960s.[37] That figure doesn't include the fertilizer market in which eight companies sell two-thirds of the 140 million tonnes purchased worldwide[38] or the proprietary seed market, where eight companies control 65 per cent of sales.[39] Just six companies—the Big Six of Monsanto, Syngenta, Dow AgroSciences, DuPont, Bayer and BASF—sell three quarters of the pesticides worldwide.[40] BigAg biotech giant Monsanto, whose tagline is 'A Sustainable Agriculture Company', spends over Rs 22 crore each day in research. For Monsanto, and their 50:50 joint Indian venture Mahyco, these are costs to be recouped and shareholders demanding a profit.

In late 2012, deep into the Punjab daily newspaper *The Tribune*, long past stories about a gang rape in Haryana and the good price that unripe citrus kinnows were fetching at market, there was a brief article with the headline, 'Danger: Groundwater Resources Plummet in Haryana, UP, R'sthan'. The author announced that India's first atlas on aquifer systems, compiled by the Central Ground Water Board (CGWB), found that aquifers in Delhi, Haryana, western Uttar Pradesh and Rajasthan, as well as the hard rock regions of the south 'have been exploited to critical levels'. The CGWB warned that 60 per cent of India's aquifers would be in a critical condition by 2025, with nearly half of those already in a semi-critical state. Rising population, erratic rainfall and slow recharge rates were to blame. There was no mention of the pressure that food and fibre crops put on water resources.

Only 2.5 per cent of the world's water is freshwater. Ninety per cent of that is used for agriculture,[41] a fivefold increase from the beginning of the twentieth century.[42] Once, monsoon rains and the five rivers that carried Himalayan meltwater down to the sea were enough to water the fields of Punjab. Not any more. The Green Revolution was fuelled not only on high-yield seeds and petrochemicals, but also on a steady supply of abundant water.

The government has been happy to help. In the 1960s, there were 11,000 tube wells in the Punjab. By early 2014, there were 1.3 million, with another 90,000 awaiting approval for construction.[43] These wells provide two-thirds of the state's water, tapping into vanishing aquifers.[44] In less than thirty years Punjab has depleted groundwater reserves that took over a century to accumulate,[45] and the electrical power needed to bring these pumps to life is provided free to farmers by the government, discouraging conservation.[46] But the cost of construction, and pumping, is growing more expensive as diggers have to reach ever deeper.[47] Crop yields rose for a time, but little thought was given to how long the aquifers could sustain such siphoning. Eighty-three per cent of Punjab is under cultivation,[48] and the pressure is intense. The National Geophysical Research Institute (NGRI) has found that every year, the water levels drop another 2 feet.[49] The bucolic rural scenery in Punjab is deceptive; beneath the surface of the fields, the water table is in free fall.

In 2012, Punjab irrigation minister J.S. Sekhon stood on the floor of the Punjab Vidhan Sabha, the Lower House of the state legislature, and stated that groundwater was continuously declining in 85 per cent of the state, while nitrate levels, caused in large part by agricultural run-off, had increased tenfold in forty years.[50] High levels of nitrates in water can cause severe health problems, sucking the oxygen out of waterways, causing algae blooms and fish kills. The same oxygen depletion occurs in human blood, causing methemoglobinemia,[51] a potentially fatal disease in infants that turns them as blue as Lord Krishna before killing them.

'Going by the excessive use of insecticides and pesticides by our farmers, [the nitrate level] is expected to reach ten milligrams per litre in the next twenty years,' junior MLA S.R. Kler, told the *Times of India.* 'This means that water in Punjab will cease to be potable for humans and animals in the next 20–25 years.'[52]

As the water table has dropped in Punjab, cancer rates have been rising, surpassing international and national averages: ninety cancer diagnoses per 1,00,000 people versus eighty per 1,00,000 nationally.[53] Eighteen people die of cancer in Punjab every day.[54] The cancers are clustered in hot spots—the cotton-producing Malwa belt, Amritsar and Bathinda—where many farmers live and work. Even with such high rates, Punjab lacks a cancer treatment facility, so every evening, at least forty and sometimes up to 100 cancer patients[55] board the train at Bathinda's railway station to travel the eight hours to the Acharya Tulsi Cancer Centre in neighbouring Bikaner, Rajasthan.[56] There, in the early morning light, taxi drivers who usually lure riders with the names of guest houses instead call out the names of doctors.[57] Patients can purchase a strip of ten diclofenac sodium tablets, a painkiller that sells for Rs 31 in the market but is provided by the Rajasthan Medical Services Corporation for a little over one rupee[58] as part of the effort to get medical services to at least one small portion of the 65 per cent of Indians who lack access to basic medical treatment. It's a pale substitute for treatment, but it helps keep the pain at bay.

Across India, there is a burgeoning grass-roots movement to lighten the heavy use of pesticides and other agrochemicals that began forty years ago. The Indian states of Sikkim and Kerala are working towards converting their states completely to organic methods, and Punjab is haltingly heading in the same direction. Andhra Pradesh is also pioneering chemical-free methods. But it is the breadbasket of Punjab that will chart India's course and it is there that some farmers are beginning to set the petrochemicals of the Green Revolution aside, transitioning to natural farming practices that Swaran Singh's grandfather would have recognized.

SPIDERS AND SWEET-SMELLING SOIL

I first arrived in Punjab before dawn, yet Umendra Dutt, director of the Kheti Virasat Mission (KVM, or Farming Heritage Mission) was there to meet me in the dark.[59] We walked through the silent streets of Jaitu, a medium-sized town of two-storey buildings, past sleeping cows and still smouldering piles of the prior day's debris and up a steep flight of stairs to the KVM office where he worked and lived. There, on either end of a hallway piled with newsletters and furled banners, were two simple rooms that serve as the base of the non-profit organization, founded in 2005, that has trained hundreds of Punjabi farmers in organic farming methods. KVM and the farmers it supports believe that the Green Revolution has failed them, ruining the land and the health of the people who work it. I had come to meet some of the farmers who argue for a return to some vestige of the old ways, when their forefathers worked their land organically, naturally, without synthetic fertilizers and genetically modified seeds. These farmers have resisted neo-liberal globalization reforms and now strive for economic independence. They are freedom fighters of food and fibre.

With the morning light, KVM's minority position in town became apparent as the shopkeepers up and down the narrow lane opened up, setting out displays of fertilizer bags and backpack chemical sprayers. Two doors down was the Bharat Fertilizers Industries Ltd. The entire street was a base for agricultural supplies, none of them organic.

Umendra, in his fifties, surrounded himself with a staff of about ten, many young energetic idealists, including Amanjot Kaur, a twenty-four-year-old woman who hailed from a family of wheat and cotton farmers near Bathinda. She earned a master's in social work in Patiala before joining KVM to lead outreach.

As the day warmed, Amanjot and I headed out to catch a bus to a nearby village called Chaina. The bus barrelled through Jaitu's narrow streets, announcing itself with an air horn that shook the brick buildings, and we climbed on board. Men sat on the left side, their heads adorned with turbans, and women on the right. Sunlight blazed through the bus windows, refracting off seats and ceiling plastered in a psychedelic green bubble pattern. We wound through tiny side streets and bumpy roads, leaving the town of Jaitu within minutes and passing through the fields that led toward Chaina.

It was the end of kharif season, the monsoon rains had passed, and the harvest was ripening by the day. There were the creamy-white flower blooms of cotton, corn stalks as tall as a man, rice paddies in varying stages, some heavy with their crop, delicate stalks bending low, others just inching up. Tractors were as common as cars on the road, which was also crowded with cycle rickshaws and camels with chins high. We passed brick kilns, the earth dug out around them, their towering chimneys like the minaret of a mosque, standing alone. White egrets and green herons poked their feet in paddy waters while hunting, songbirds were notable only for their absence. Studies have shown a direct link between the use of neonicotinoid insecticides and declines in birds as well as important pollinators such as honeybees.[60]

After a fifteen-minute ride, Amanjot and I stepped off the bus and walked down a chalky road bordered by rice fields; we were bound for the farm of Sukhdev Singh, who goes by the nickname Gora, one of the KVM farmers.[61] On the way to his patch of land, empty bags that once held pesticides lay spent on the path. The chugging of a bore well pump shook out from a shed and water bubbled into a canal. Across a field a figure waved, and we balanced along the berm of the narrow canal to reach him.

Gora met us with a smile. In his early thirties, he wore a long, loose white button-down shirt and salwars. His cheeks were dark from the sun, his beard was trimmed, and his thick straight hair stuck up at odd rakish angles that a city boy might work very hard to achieve.

Gora didn't let go of his daati, a small sickle-shaped tool with a sharp blade, as we sat in the dirt on the edge of his 5-acre farm. The implement was an extension of his body. It was a digging device and slicing mechanism, a pointer and harvester. Nearby, a labourer he had hired to help with the mung bean harvest continued to add to his pile of small plants, each with a handful of bean pods clinging to them.

Gora told us that he switched to natural farming because of his late mother. Four years earlier, she was suffering from pulmonary problems and asked Gora to stop using the pesticides that she believed were making her sick. He had always used chemicals, he said, because he didn't know that you couldn't. Following his mother's advice, he used organic methods on 2 of their 5 acres, growing food for his wife, their eleven-year-old son, and his brother and his brother's family. On the other 3 acres, he still used pesticides, he said, but less each year, improving the tilth of the soil in preparation for going completely organic.

KVM helped along the way, showing him how to plant mung beans, also known as green gram, a natural nitrogen fertilizer, at the feet of the corn plants. How to use native seeds, not hybrids, so that he could save them year after year. How to rotate and combine crops of corn, sorghum, wheat, sugar cane, mung beans and vegetables, so that his family could grow on their land almost everything they needed. Now, they could make brown sugar from their own sugar cane and draw mustard oil from their own mustard seeds. There was enough fodder for the buffalo, cows and an ox, and the surplus went to market. Gora learned about seed selection, seed treatment, sowing techniques, mixed cropping, and how to make bio-pesticides by cooking neem leaves and other plants in cow urine. He composted field and household kitchen waste and added it to the soil.

Like his mother, Gora came to believe there was a health benefit to farming without chemicals. 'And it tastes better,' he said, poking at the earth with his daati. He smiled like he was in on a secret.

But the biggest change, he said, was in the soil. He pressed harder into the daati handle and lifted up a handful of earth, then crumbled the soil and held it to his nose.

'Organic soil smells so good,' he said, holding it out to me like an offering, a circle of red henna revealed at the centre of his open palm. I held the handful of earth to my own nostrils, so close that a clump stuck to the end of my nose. It did smell sweet, almost edible. Like a wild mushroom. Alive. Though the surface of the ground where we sat was cracked, the dirt was laced with the scent of petrichor, the aroma of rain on dry earth. The word combines the Greek *petros,* meaning stone, with *ichor,* the fluid that runs through the veins of Greek gods. During dry periods, plants exude oils that permeate the soil they inhabit. When moisture arrives, the bouquet is released along with geosmin, a metallic by-product of the natural actinobacteria that live in soil and play a key role in decomposing matter and building healthy humus. (Actinobacterias were also the first antibiotic cure for tuberculosis.)[62] Without pesticides, the soil was more complex and could retain water for longer. It was richer, darker and heavier, containing a living history of weather, wind, sunlight, plant redolence and microscopic bacteria.

Herbicides and pesticides kill indiscriminately, destroying a harmful caterpillar but also the pollinating butterfly it might have become, along with other beneficial insects that lived alongside it. In the earth, pesticides, herbicides and fungicides break down natural bacteria and fungi, washing the dark colour from the soil, while dung and other organic materials add to it, supporting an ecosystem that surrounds and supports plant roots.

Gora said his wheat yield dipped at first, but then recovered to pre-organic levels.

'I'm not worried,' he said. 'Each year the yield goes up. With chemical farming, work is only done on the crop, not on the soil.' Gora's goal, like every organic farmer, is to improve soil conditions over time by providing a strong foundation. Those little shin-high mung bean plants growing among the corn made more than a bean harvest. Gora reached for one of the plants and pulled it up by the roots, shaking it free of the sweet soil. He held it in his hennaed hand and pointed to little nodules along its root system. Mung beans and other plants in the legume family *Fabaceae*—from clover and peas to acacia trees—are nitrogen-fixers. They use photosynthesis to snatch nitrogen from the air and 'fix' it, chemically transforming it into a form usable to plants via little nitrogen pellets the size and colour of tapioca pearls. These nitrogen-fixers work much like urea, though in a slower, steadier way.

Nitrogen, along with potassium and phosphorus—NPK in agricultural shorthand—are the three essential building blocks of plant life, working along with other macronutrients such as calcium, magnesium and sulphur. A host of micronutrients such as copper, iron and zinc play their important supporting roles, just as they do in the human body. As his soil developed, Gora said, his expenses dropped, since he bought less and less from off the farm. But without pesticides, what about the pests?

Gora leaned into a stalk of corn and pointed to a black and yellow spider. In its clutches was the remainder of a grey bug. Before he switched to natural farming, he said, there were no insects, good or bad. Now insects abound. Everywhere we looked we saw spiders, all different varieties, tucked into the rigid folds of striated corn leaves. With each step I took on the cracked earth, black and white spiders scurried away, disappearing into the fissures. They were everywhere. Three different types of butterflies passed by.

In Chaina, so close to KVM headquarters, about eighty of two hundred families have begun to farm naturally. Those who could return only some of their land to organic chose to reserve that harvest for their own families, sending the conventionally grown chemical crop to market. But the natural farmers are still a tiny fraction of the farmers in Punjab and across India.

I pressed Gora. Surely it isn't that easy. If it was, everyone would leap back to organic methods.

'They're used to chemical farming, and they don't want to take risks,' he said. 'And, it's more work!' What the chemicals don't do, humans must. Natural farmers spend more time in their fields, tending and weeding, and more time over a hearth, boiling down dung, urine and herbs to make jeevamruta, the traditional Indian fertilizer. Pesticides are made from the leaves of neem, camphor, marigold and datura, and the five-leaved chaste tree, as well as garlic, onion and soap.

Gora Singh looked pleased. He smiled, sitting in the shelter of his corn stalks, poking at the dirt, piles of mung beans beside him, spiders hidden about, the sweet smell of his soil lingering in his nose. The greatest reason to change, he said, was 'a love of nature'. This was his form of Swaraj, Mahatma Gandhi's idea of self-rule, of independence and freedom. He was no longer dependent on the bags of petrochemicals that lined the street outside KVM's office and every street like it across all of South Asia.

BUST AND BOOM AND BUST AGAIN

Down the road from Gora Singh, Amarjeet Sharma tended his own plot of land.[63] Where Gora walked light upon the earth, Amarjeet seemed heavy with sorrow. He had been farming his 4-acre plot in Chaina for thirty-six years, growing cotton, sugar cane, fodder for his cows and vegetable staples like chilli and bhindi since he was a nineteen-year-old boy. Seven years earlier, he had switched to organic.

We sat on a high table—a place for work and afternoon rest—underneath a lone tree that sprinkled small golden leaves down on us as the breeze blew. They alighted on his pale chequered turban, on his greying beard, on the table between us.

'We used to be prosperous,' Amarjeet began. 'From 1970 to 1991, I was very prosperous. It was the Green Revolution boom. Now it's switched. Families won't even come visit their farmer relatives for fear the farmer will ask for money.' In 2000, Punjab lost its status as the Indian state with the highest per capita income, relinquishing the title it had held for four decades, correlating with the peak years of the Green Revolution. Now, nearly two-thirds of Punjabi farmers hold more debt than the national average.[64]

'From 1991 to 2000, I had a radical loss in cotton yield,' Amarjeet said. By 1992, the yields of the supposedly high-yield crops they'd been sold on a generation earlier were flatlining. It was the same all across India. By 2004, per capita food grains production had dropped back to 1970s levels.[65] But there was always something new on offer.

As yields began to wane, another high-tech improvement arrived from Monsanto in the form of genetically modified Bt cotton seed. Bt is short for *Bacillus thuringiensis,* a natural bacterium that is harmful to certain insects, especially the larvae of many moths and butterflies, including the bollworm, a cotton grower's nemesis. Monsanto took the genetic coding for the Bt toxin and spliced it into the genetic make-up of the cotton seed. The hope was that the built-in resistance would allow farmers to use fewer pesticides. In 2002, the Indian government approved the use of Bt-modified cotton,[66] and four years later Bt cotton covered 3.8 million hectares, more than a third of India's cotton-growing territory.[67] It took less than a decade for Bt cotton to corner 90 per cent of the country's cotton market. Yields spiked, and then—gravity as surely—ecology, biological adaption and water scarcity pulled them down again.

In 2006, Cornell University researchers studied the longer-term economic impact of Bt cotton.[68] The first three years were great. Chinese farmers in the survey who had planted Bt cotton cut pesticide use by more than 70 per cent and had earnings 36 per cent higher than farmers planting conventional cotton. But seven years after the initial planting, they had to spray just as much as farmers using non-Bt cotton, up to twenty times per season. That resulted in a net average income of 8 per cent less than conventional cotton farmers because Bt seed is triple the cost of conventional seed. Bt cotton also did nothing to protect against fungal attacks or what are known as the 'sucking pests': aphids, stinkbugs, thrips, spider mites and mirids. Given this mixed result, the introduction of Bt cotton, as the first transgenic crop allowed into India, has been hugely controversial. Was it a miracle crop doubling cotton yields or a debt-inducer driving farmers to suicide? Or both?

For Amarjeet and his fellow farmers, Bt cotton and other agribusiness products and methods were tickets to a roller-coaster ride with a steep annual admission price. By 2005, Amarjeet wanted off. He recalled the days before the agribusiness carnival. The cotton they grew was rain-fed, and the yield was a third of what the irrigated cotton produced, but part of that increase, he explained, was continuous planting. Before, in the seasons without rain, you'd let the field lie fallow or just grow vegetables or a lighter crop. Now the fields are never dormant, so of course the yields have risen. During the Green Revolution, if you expected an increase of ten quintals, you got twelve. Now, Amarjeet said, it was always less than you expected. So Amarjeet returned to planting local varieties of cotton from Punjab and Rajasthan, which had evolved with the region's climate and soils over centuries. 'Now,' he said as he sorted through small piles of seeds wrapped in the folds of newspaper scraps, 'I don't have to buy seeds.'

Farming has always involved risks. Monsoon rains come, or don't. There are years that are too hot or too cold. But the promises of agribusiness worsened the cycles of crippling indebtedness. Farmers found themselves going deep into debt at the start of each season, and then doing anything and everything they could to ensure that the crop would pay off, come harvest. The hope for pay-off was so desperate and strong that they'd chase it with more money, more fertilizer, more pesticides, wagering that they would recoup their losses, come harvest. Unlike Monsanto, they had no shareholder to shoulder the burden if the gamble didn't pay off.

India is a land with six hundred million women and men who labour on small farms; only a tiny 1 per cent of all farms are larger than 25 acres. Like those of Amarjeet Sharma and Gora Singh, the average farm is about 3 acres in size.[69] Farmlands abut one another like a patchwork quilt composed of remnants gathered from a seamstress's floor. Aerial views reveal a mixture of elongated rectangles, a rhombus here, a trapezoid there. We sat perched on Amarjeet's tree-canopied table at the centre of his few acres. A hundred feet in one direction, women and children picked cotton on a neighbour's land. A hundred feet in another direction, two men walked by with pesticide sprayers on their backs—'nozzleheads'[70] in Western lingo—each rhythmically pumping down on a hand lever that spumed pesticides out in a mist they walked through and inhaled. It was peaceful as they moved apace, a dozen feet from each other, walking up the rows and turning to return down the field. There was the soft pumping sound, the spray landing like a magical mist on the cotton or drifting (some of it) on to Amarjeet's land. Tiny farms like Amarjeet's that chose to forgo chemicals or genetically modified crops would always have to contend with drift and cross-pollination from their neighbours' land.

Amarjeet echoed what I heard many farmers say: labour was their greatest problem. 'No one wants to come and work on my farm because it's too much work. Pesticides are like intoxicants—they're addictive. And it's more convenient to spray.' He was describing local labourers' preference for working on conventional farms, but I also heard about a general labour shortage in Punjab. The wealthy state has long been dependent on migrant workers coming from poorer regions of India, the equivalent of the Latino labourers who keep strawberries and lettuce on American dinner tables. As the economies of some states—such as Uttar Pradesh, Bihar and Haryana—improved, the men stayed home to work, leaving Punjab with a shortage.

But even with the lack of available help, Amarjeet remained determined. 'Small farmers like me are experimenting with farming naturally, but I do this out of goodwill. I do this like Bhagat Singh,' he says, referring to the freedom fighter.

For Amarjeet, to turn to natural farming was to assert his independence. Back in the 1970s, he married his wife in a nearby school and enrolled as a member of the Communist Party of India (CPI). Agribusiness, he said, 'is a mental and intellectual and ecological colonization'. The wind stirred and more golden leaves cascaded from above. 'The Green Revolution brought so many changes to farming. Not just the chemicals, but the whole entry of commercial banking, of tractors and diesel and banks and technology.'

He could have been channelling the Kentucky farmer Wendell Berry, who has written extensively on the relationships between humans and the land they inhabit. In his short but seminal essay 'What Are People For?' Berry wrote, 'Since World War II, the governing agricultural doctrine in government offices, universities and corporations has been that "there are too many people on the farm".' The goal has been to replace humans with 'machinery, petroleum, chemicals, credit and other expensive goods and services from the agribusiness economy'. Though working under the noble guise of feeding humanity, this economy is enormous, swallowing the small farmers who step into it, and forcing many off their farmlands altogether. The great question, wrote Berry, is 'What are people for?'

'It is apparently easy to say that there are too many farmers, if one is not a farmer,' Wendell Berry wrote. 'This is not a pronouncement often heard in farming communities … No agricultural economist has yet perceived that there are too many agricultural economists.'[71]

Amarjeet did not see himself as a surplus farmer.

'I see the farm as a living being,' he said. 'Nature tells me what it needs. I don't need scientific universities.' He wanted a more balanced approach that minimized the vagaries of the agribusiness model and worked within the rhythms of nature.

'The Green Revolution sucked the life from the soil,' he said. 'You can't bring it back to life with manufactured supplements.'

Amarjeet said he once heard the former prime minister Manmohan Singh, an economist, say that money does not grow on trees. Amarjeet objected. 'If it doesn't grow on trees,' he said, 'then where does it come from? If you don't have trees, what do you have? There used to be more trees, but they've all been cut down.'

It was later that I learned that Amarjeet's only son had died six months before from a bad batch of moonshine. Perhaps he had been one of the 45 per cent of Indian farmers who want to quit farming, as revealed in a national survey.[72] Amarjeet's aged body sat on the table, picking out the best seeds to plant the next season. But who would plant them with his son gone? Other farmers' sons were leaving too, in pursuit of education or promises of steadier work in the cities.

We parted quietly. On my way back to the road, I stopped to speak to the labourers who had been picking cotton near us. The matriarch of the family that was packing up at the end of their workday was not yet forty and was accompanied by her daughter, her two sons and their wives and babies. Together, they worked from 8 a.m. to 6 p.m. every day, she said, with each family member earning a hundred rupees, more or less. To buy an acre of Punjabi farmland today can cost sixty lakh rupees.[73] Even if all eight family members of their family workforce, including the baby, worked every day and pooled every rupee into a land fund, it would take them hundreds of years to purchase a single acre.

How Amarjeet would have recoiled had he seen the agribusiness of the West—the massive farms run by computers, humans absent from the landscape. If he fears that this approach might one day render him obsolete, it is a well-grounded worry. But India is still a long way off from that scenario. While factory farming might be weeding out small family farms of America, small landholders with just a few acres still dominate global agriculture. A large landholder by Indian standards is someone with just a few dozen acres. Someone like Vinod Jyani.

OVERNIGHT

Vinod Jyani was just a baby when spray planes and carts of free petrochemicals showed up at the farm just a few miles from the

Pakistan border in the Punjabi village of Katehra in Fazilka district. Growing crops with chemicals was all he knew until the fall of 2005, when Vinod went to a meeting 'to oblige a friend' and heard Umendra Dutt of KVM speak about organic farming. His response was akin to a religious conversion.

'That was it,' he told me as doves cooed from the eaves of his house, a sprawling complex set amid the 130 acres that has been his family's farm for seven generations. It was like a 'light went off'.[74] He immediately bought Subhash Palekar's book *Philosophy of Spiritual Farming,* which speaks of the birthright of every human to 'poisonless food, pollution free water, air, environment and happy, pleasant, wealthy and prosperous life'. The problem was the Green Revolution: 'violent, nonscientific, inhuman, barbarous, demonous, atrocious, ferocious, monstrous'. The solution was natural farming, a 'legal, non-violent and constructive, non-political, non-religious movement'.[75]

Vinod devoured Palekar's writing and two weeks later attended a workshop organized by KVM.

'The very next day, on 20 November 2005, I took all chemicals from my farm,' he said. 'I started with a passion—and a zero budget.' He was smiling as he sat at the centre of his now successful organic farm, but when I asked him how the transition went, he laughed. 'It went bad,' he said, shaking his head. 'Bad! For three years it was a struggle, but I was committed.' He was in his early forties. There was time to adjust to the change, and he had resources to cushion the transition.

'I was lucky,' he said. 'I had no doubts. I'm the only farmer that went full-fledged in one go. Yields didn't go up in all the crops right away—mustard, millet, sugar cane, citrus—but they went up after four years. Wheat has a maximum yield of only 14 quintals. We're doing 11 now.' He was sold.

As we spoke, he sat with his wife and mother and daughter

in the courtyard of their house, the moon on the wane and Cassiopeia hovering in a star-littered sky overhead. There was no television or radio, just a small electric fan to keep the mosquitoes at bay. A tulsi (holy basil) plant was growing in a pot, and the evening's worship lamp glowed beside it. More servants than I could keep track of cleared our plates after an exquisite but simple home-grown meal of vegetable sabzi and fresh rotis, the smell of the fire searing into the freshly ground wheat. All the cooking was done over a large open fire, which Vinod's wife insisted caused no pollution. 'As long as you do the proper rituals with milk and ghee, make offerings, give the first chapati to the cow,' she told me, 'the smoke is not a problem.'

Their daughter Isha reclined on a charpoy, wearing jeans and a little denim vest, cracking her knuckles and pulling her fitted red T-shirt down each time it shifted up her torso. She was on fall break from Delhi University, where she was studying philosophy. Back home, she was bored and told me she missed the shopping.

Vinod's wife Indra was a regal woman, with perfect eyebrows, a pink salwar and shoulder-length hair. She was active in politics in the ladies' wing of the Congress party and fully involved in the running of the farm. I asked her what she thought when her husband made this huge change overnight. Caught off-guard, she blushed.

'No one has asked that,' she said, at first deflecting the question with a laugh before answering. 'His family didn't understand. 'You're highly educated,' they said. So we went to our temple and the priest asked me, 'What are your duties? Not just the money, not just what type of wheat, but what are your duties? What kind of food do you give to everybody? Beyond family, what is your duty to something higher?' She decided to support her husband's decision.

While Vinod farmed the land, Indra ran the retail business,

creating value-added products—fresh juices and honey and packages of dried beans—that helped keep the operation economically viable.

'But it was up to my husband,' she continued. 'Indian ladies, Indian wives, we are not interfering. But ask his mother. Mothers interfere!'

So I asked Vinod's mother, an elegant old woman with her dupatta pulled over her head, framing her face in lace.

'I tried to talk him out of it,' Vinod's mother Indu said. 'Before, we had a lot of wheat, now the yields have gone down. But I remember the spray planes and the helicopters when the Green Revolution began. We had two hundred acres of wheat. This house was the storeroom!' she said, gesturing to the sprawling building around us, and then looking at me directly. 'People didn't get sick.'

'The side effects came later,' Indra countered quietly. 'The land degradation first and now Bathinda has the highest cancer rates. This whole area is known for cancer.'

'Oh,' said the mother, 'but our yield was so much more then.'

Vinod Jyani was a man who got things done. We were up at dawn, meeting by the small pool to do yoga. He prompted me to lead the vinyasas but asked me to await the moment the sun actually tipped over the horizon before we began the sun salutation. It was perfect. Afterwards, we headed away from the buildings to tour the farm's acreage.

The land demanded a staff of seventeen who milked the dairy cows, monitored the methane digester to make fuel from manure, and tended the crops and orchards. Ducks and geese roamed freely, and visitors came from Spain, Argentina, Australia and all over India as ecotourists. WWOOFers, volunteers who worked in exchange for room and board through World Wide Opportunities on Organic Farms, added to the labour pool. Vinod sat on the board of KVM, was involved with local politics, and was interrupted frequently throughout the day by men asking for favours.

Farms with such diverse resources and income streams were beyond the reach of the other 99.9 per cent, but they existed. Vinod knew his relatively large landholdings gave him an advantage. It allowed him the space for diversity, both physical and economical.

'My father was a pioneer of the Green Revolution,' he said as we walked in the early morning light.

'The Indian government, along with the American companies, sponsored aerial spraying for three years. The farmers became dependent on it,' Vinod recalled. 'For those first three years, the plants did better. Everyone was addicted. Then the government stopped the spraying and gave out spray pumps and subsidies to buy the chemicals. Use this, they told the farmers, and get a better result.'

Next came improved seeds from the universities. Then, in the early 1990s, hybrids started to appear. A decade later, Bt cotton seeds arrived. But they still needed pesticides. Revenues went up, but so did expenditures. An economy based on chemical farming emerged that reached from the Missouri headquarters of Monsanto to Mahyco in Maharashtra to the tiniest seed shop next door to KVM. Vinod Jyani, Gora Singh and KVM, though each dwarfed by the scale of multinationals or even the average American-sized farm, posed a threat to that system.

'Farmers growing chemically get subsidies,' Vinod said as we approached a field of millet. 'I don't get any subsidies.' There are 82,000 paid agricultural scientists in all of India, he told me, and 99 per cent of them are doing chemical farming research. There is no one to lead research for organic farming. It was a complaint Rachel Carson had made half a century earlier, when she reported in *Silent Spring* that 98 per cent of American economic entomologists in 1960 were researching chemical insecticides.

A flock of sparrows passed overhead. Hardly seen in rural areas any more, they had all but vanished from Delhi in recent years. Vinod figured the birds ate a small percentage of his crops, and he let them. He had less tolerance for parakeets, a group of which had arrived with furious squalls before settling over his fields; they wrought havoc on his crops. Nearby, an older man with dark sunglasses and a stunning white mustache emitted a loud whoop into the air. The parakeets lifted and circled, and I jumped when someone out of sight fired a blank, sending a boom into the still morning air. The birds scattered in a sweep of green across the sky, leaving behind the millet, ripe and heavy on pendulous stalks.

Close to the road that bordered one side of the farm, Vinod had planted a buffer of trees: shading jujube and flowering bougainvillea, thorny shrubs as a safe haven for small birds to nest, taller rosewood trees for predatory birds to perch, and others leguminous species whose task was to fix nitrogen with their roots. Someday it will grow into a small forest providing shade and protecting the area from winds. We looked across the street to the next farm, an expansive monocrop of Bt cotton, uniform and empty. It looked quite boring in comparison. 'No birds. No trees,' said Vinod. 'I feel really sorry for this guy, ya? He'll get 50 per cent of my yield.'

'You have to study the nature,' he said, tilting his head up. 'Listen, can you hear the birds talking?'

We walked in silence, listening, as we headed to a side entrance to the farm, where a guard sat by a small tent. He was a sentry to the sweet lime trees, on the verge of harvest, protecting them from human and animal scavengers. Honeybee boxes stood at attention nearby, the insects passing in and out industriously. Among the trees were rows of cowpeas that fixed nitrogen for the lime roots to absorb. After the harvest, cows could graze on the plants.

'Chemical farming is asking the land to do something it wasn't made to do,' Vinod said. 'If you work sixteen hours instead of eight, you'll have to rest sooner. This,' he said, gesturing to the agricultural ecosystem we stood amid, 'is all an integrated system. I have so much more time now than before. When I had a chemical farm, right now I'd be at market, buying things. Now I have more time with my family, and to spend walking the farm and thinking about what I can do.'

Vinod believed that India, like him, could switch to organic overnight, starting with his own state.

'All of Punjab can become organic in one single day. One day. One single day. India will not go hungry,' he said, without an ounce of doubt.

'In India, population is our strength, but everyone is taking it as a problem. What if…' he began excitedly. 'What if someone comes up to you and says, "Meera, you're so ugly." And then another person comes and also says, "Oh, you're so-o-o ugly." And then another? It is like this. Other countries say that our population is overburdened. They say we're weak. They sell this idea to the government in five-star hotels with AC.'

Vinod stopped and turned to me, 'India's population is her strength,' he said. He spoke of how all those hands could help work the land without chemicals. His eyes turned back to the trees, where a sparrow hopped along under young sweet limes. 'Think what we can do.'

FIND THE WAY

Some of the farmers I met found inspiration from Gandhi. Others from spiritual gurus. One, from Bob Dylan. In the home of Ashish Ahuja in Abohar, Punjab, there was a poster of the young musician declaring, 'The times they are a-changing' hanging above a shelf lined with books: *Plant Science, The*

Complete Book of Composting, The Handbook of Organic Gardening. Young Indians like Ashish were as enthralled with the idea of growing food without chemicals as prior generations had been about growing with them. An urban boy who'd transformed himself into a born-again farmer, Ashish had grown up in a Delhi suburb, though his neighbourhood of Vasant Vihar was still wild enough that a friend had had a tree house. For three months of every year though, he roamed his family's land in Abohar with his cousins, and the big sky in the countryside imprinted on his soul.[76]

Like most upwardly mobile families, Ashish carried through with expectations to pursue a professional life. First he studied chemical engineering in Delhi. Then he went to a graduate school in the United States in the mid-1990s. He was giddy to be in the land of Dave Matthews, the Counting Crows and the Grateful Dead, but while his friends went off to schools like University of Massachusetts in Amherst, he ended up in Lubbock, Texas, a place not so different from Punjab, where people were toughened by exposure to hot sun and thick dust. He returned to Delhi and settled down to an engineering job, but as his thirtieth birthday approached he began to fantasize about those childhood stretches in Punjab, and he imagined working on the land instead of with computers, of growing his own food without fertilizers and pesticides.

His return to India coincided with Vandana Shiva hitting the international circuit railing against agribusiness, promoting organic farming and seed saving, and fighting corporate American attempts to patent neem, an Indian plant long used for its medicinal properties. This was the era of Operation Cremate Monsanto,[77] in which genetically modified seed fields were set ablaze in protests that combined nationalist appeals, opposition to rampant globalization and rejection of genetic engineering. Going organic was a movement, and albeit in his own quiet way, Ashish wanted to be part of it.

During the next decade, Ashish met his wife on Shaadi.com. She designed textiles at Auroville in Tamil Nadu, and was open to the adventure of a farming life. Along with their two young children, who vied for their allotted hour on the iPad, they lived in an apartment in Abohar as they prepared to build a home on the 90 acres of family farmland that he'd taken over. Ashish had already transitioned 25 acres to organic.

His father was delighted with his son's move. He had wanted to farm himself but had succumbed to the maxim that only the most useless son farmed while the smartest children became businessmen or doctors. Ashish's mother was less enthusiastic, and his cousins, who managed their farmlands remotely and only visited when they were in the mood for a picnic, were baffled that Ashish actually wanted to live in the country. His educated friends were jealous, imagining days beyond a cubicle.

Of course, Ashish had no idea what he was doing, and more than a decade into it he still felt like he was figuring it out. After dropping off his younger child at nursery school, we drove the 18 miles from the Abohar apartment to his farm, passing rice paddies and cotton fields and acacias in full bloom. We passed bright green Balkar threshers and jugaad vehicles, Mad Max mobiles pieced together with available body parts and powered by diesel engines designed as irrigation pumps.

'Everything in Punjab is about the American model. The locals follow the US example. Americans don't need to label GM (genetically modified) foods, so why should the Indians?' he said, as the Counting Crows belted 'You Ain't Going Nowhere' from the CD player of his Scorpio SUV. 'But at the same time, they'll use the reverse logic for something else—oh, we're Indian, we can do it our own way!' These duelling polarities alternately lionize and demonize the West.

In the village of Medikhere, past a mud wall, Ashish pulled down a side road and parked under the shade of a lone tree at the edge of some of his cotton fields. 'The women like the cotton harvest,' Ashish said, slipping on a sun hat and casting his boyishly handsome face into a shadow. 'They are alone in the fields with each other, and it's a time to gossip and talk.' We walked towards the soft sound of women's voices, their bodies hidden by the six-foot-high plants. Some flowers were still setting, their petals pink and white. Cloudbursts of cotton exploded from seed heads. The women would pick cotton for about six hours of an eight-hour day, with time for lunch and tea breaks. They earned six rupees per two pounds of cotton. According to Ashish, the best of the harvesters could gather sixty-six pounds in a day's work, meaning they could earn much more than many day labourers.

The women's voices hushed, and they swept their dupattas completely over their faces as we approached. When Ashish asked them to show me how they harvested, their fingers flew across the bushes, like bees pollinating, the reaping tossed into a sling across their backs. I reached for one cotton bloom, unprepared for the sharp points where the seed head opened in five folds, digging into my fingertips. I yanked awkwardly at the soft cotton, and they laughed, smiles seeping through the gauzy fabric. That one's not ready, they explained, guiding my hand to a better bloom.

Ashish pointed out a blue shed on the main road at the edge of the fields. It housed a reverse osmosis water filter that the government installed to clear the local well water of contaminants such as nitrates and pesticides that percolate down from treated fields at the surface. But as the wells were dug deeper and deeper—some 400-feet down—they tapped naturally occurring uranium, which releases radon gas and arsenic. Both have been linked to cancer, and the filtering doesn't remove radioactivity. 'What to do?' Ashish said. The canal water was for crops, and the well water was too saline for farming, so the at-least-partly filtered water served the people's drinking needs, contaminated or not. A survey of three thousand village wells around India revealed that a fifth of them contained nitrates in excess of WHO limits.[78]

Next we went down the road to Ashish's main farm, where he grew wheat, sorghum, black gram, mung beans, channa and vegetables. There were acres of kinnow and more cotton, both Bt and non-Bt varieties. A small array of solar panels pumped canal water into a large pond.

'I'm not making any money off the organic,' Ashish said as we walked through the fields, admitting that he was unwilling to wrestle with the certification paperwork that might grant him a higher price for organic crops, even in India's limited market.

He pointed to where his house would someday sit, envisioning a time when he could walk out of his front door in the morning, coffee cup in hand, to oversee his farm. But first he wanted to upgrade the housing for his five full-time staff and their families. I asked him what his staff thought about his attempts at organic. 'They're probably laughing at me behind my back,' he said.

Ashish wasn't involved with KVM or any other activists, but Dr Rupela, former principal scientist of International Crops Research Institute for the Semi-Arid Tropics (ICRISAT), had come by the farm. Rupela noted things Ashish could do to further improve the organic outputs of his land. For example, Ashish hadn't yet dug trenches around each kinnow tree and applied jeevamruta. That might help combat the biggest kinnow pest: citrus psylla. He was using a few pheromone traps that lured pests on to sticky board or through one-way chutes into bags that dangled on poles. We crossed a path, from the organic kinnow to the conventionally grown, and he pointed out that trees were stronger when the chemicals were used. But with both the organic and conventional, the kinnow gave him such trouble that he was tempted to replace them all with guava.

'When I was growing non-Bt cotton, I would be using twelve pesticides—broad-spectrum organophosphates—the worst ones,' he said as we passed into more fields of cotton, butterflies coasting through the air. 'But with Bt, I use maybe four or five sprays.' He'd given safety gear to his workers, but it was too hot and they took it off. 'They feel like they're suffocating,' he said. Bt cotton, he said, only worked where there was water, because it needed so much of it.

'Local non-Bt cotton outperforms by a mile,' he said of a variety he got in Rajasthan.

What really mattered, he found, was what had been planted in the field before the cotton. 'The non-Bt cotton I grew on a former legume field was twice the size of the one I grew on a wheat field.'

As we circled back towards the front gate, Ashish said he was 'totally committed' to switching to organic.

But what about the rest of India? 'If we can get yields up, then there is a possibility, but I can't think about that,' he said, adding, 'I'm just trying to figure out my one farm.' Still, it made sense to him that organics should work. 'The average farm puts 250 kilograms of urea on each acre,' he said. 'And 100 kilograms of DAP,' meaning diamonium phosphate, another chemical fertilizer. 'And another 50 kilograms of potash. That's 400 kilograms of fertilizers per acre, half of which is filler and who-knows-what. Potash has naturally occurring uranium in it. It's from the earth. There was a study that found radioactivity in children's hair.' He shook his head.

Eliminating all these inputs should help balance the ledger, even if the yield was lower, so Ashish was hopeful. 'The government has to support it,' Ashish said. 'If we can improve the crops through technical support for organic, the economics will work out. Wherever the free market has come in, it doesn't work.' It might work perfectly on paper, but there wasn't enough consideration of local regions and variations of impact.

'The main thing is we have great brains, but the priority has to be different,' he said. 'They're all working on chemical farming now, but if we could get the scientific and technical support for organic, this could work. It would get more farmers interested in it—real farmers, not like me, who is afraid to drive a tractor!'

But what about the real farmers who know how to drive tractors and the ones too poor to even have an ox to lead a plow? Ashish had large acreage, savings, an education and ambition. He was able to take risks unthinkable to millions of farmers. Could 'real farmers' make such a leap?

Back in the car, we returned to Abohar. The CD had shuffled to Eddie Vedder's *Into the Wild* album.

'I won't be the last, I won't be the first,' drifted the husky voice from the speakers, 'Find the way, to where the sky meets the earth.'

YOUTUBE FARMING

Far from Punjab, I later stood on a corner in a place known as Electronic City on the outskirts of the southern city of Bengalaru (formerly Bangalore). The city's rapid growth during the information technology (IT) boom made it the Silicon Valley of India, home of such IT giants as Infosys and Wipro, and development had spilled over into the once quiet countryside. When I arrived in 2013, Electronic City felt like it had fallen from the sky and the dust was still settling. Building cranes dominated the emerging skyline, adding layers to freshly poured concrete buildings surrounded by bamboo scaffolding. Cows trotted down the street, startled by the passing of corporate buses where there had once been bullock carts.

My cousin Pavitra was travelling with me, and we were bound for an organic farm of a very different sort than what I'd seen in Punjab, one that reflected a new India as buzzing as Bengalaru. A silver Tata pickup truck stopped in front of us and Nameet MVS, who went simply by Nameet, stepped out to greet us.[79] The owner of FirstAgro and a first-generation farmer, he instantly reminded me of Ashish Ahuja. He was tall and fit, a man who had left India for North America, where he flew as a pilot in Canada for eight years, and then had returned to India to reinvent himself as a modern farmer. He wore a white-collared T-shirt, periwinkle shorts, hiking boots, and a crooked smile with a broken front tooth. At thirty-five years old, Nameet had ambitions that far exceeded those of coffee-cup-toting Ashish. Nameet wanted to take what he called 'Zero Pesticides', certified to international standards, gourmet and epicurean, to scale.

FirstAgro's objective was to go big and do it fast, without any agrochemicals. Nameet and his brother Naveen founded the company to bring high-end organic produce and fruit to the growing Indian elite market. Raised in an army family, the boys had lived in Singapore, Japan, Dubai, San Francisco and all over India while they were growing up. Naveen still lived overseas, in Japan, while Nameet ran the farm. In 2011, they bought 45 acres of bare land in Talakadu, in the Mysore district of Karnataka, and by 2013, 22 acres were under cultivation. By selling one tonne of produce daily, Nameet said, they were earning over Rs 6.8 crore each year. Soon they hoped to recoup the eight crore rupees they had invested, and there were plans to buy more land and expand operations. One brother hoped to run sixteen different farms around the country; the other sought to start a few larger farms.

'I don't know if this is going to work,' Nameet said. 'I figure, either way, we'll be a textbook case. Either this is how you do it. Or this is how you don't.'

In the truck with us on the three-hour ride from Bangalore to the farm was Rajesh Nayak, who worked at Alpine Vineyards, a nearby state-of-the-art winery that, if their branding took, would stand with FirstAgro as a Cauvery Valley ecotourism destination. Together, they envisioned the birth of an Indian artisanal food movement to rival the greenest pockets of Italy and Vermont.

They were off to a good start. Nameet was returning to the farm from Bengalaru after a meeting with the chef at the Ritz-Carlton to design the latest menu. His other clients included the Taj, Marriot and Hyatt hotels as well as brightly lit, air-conditioned grocery stores such as HyperMarket and Nilgiri, which are patronized by housewives who watch cooking shows, he said, and many, like himself, who had brought back a new set of tastes from time spent abroad.

We left the well-paved, expansive highway that radiated from the city, pausing to pay at a tollbooth. Nameet sniffed the air. 'Malathion,' he said. 'One of these trucks is carrying pesticide. You can smell it.'

Off the highway we travelled into Indian countryside. We passed a string of villages with fields in between. Finger millet stalks with puffy fronds hovered close to harvest. Sugar cane with tall tassels reached high, and young rice paddy swam low. Coconut trees poked out of a rocky landscape. In one village, an old man walked by a chai shop, inexplicably dragging a dead puppy on a rope.

There was no grand entrance to the farm. Nameet turned off the road onto scrubland tinged red with iron-rich soil, and we drove through brush. Mountains came into view and rock formations rose close by like Devil's Tower. Nameet paused the truck, gazing at the hilly horizon, and said this is where he wanted to be, 'eating cherry tomatoes and drinking Shiraz, the mountains in the background.' Maybe he wasn't so different from Ashish Ahuja after all. He broke into Louis Armstrong's '*It's a Wonderful World*' as he pressed on the accelerator and we bounced along the road to arrive at a small building next to a cloth-covered greenhouse. There was a table of mixed chillies drying in the sun. A dog bounded into his arms. Nameet had just acquired five shepherd pups from a church, bringing FirstAgro's canine population to eleven. Up in Delhi at his in-laws house, his wife, an urban designer from Punjab whom he met at a party in Canada, was just weeks away from delivering their first child.

Nameet did not hail from a lineage of agriculturalists. He learned to farm, he said, from '800 hours of YouTube education'. He passed the e-knowledge he gleaned to his workers each time he circled the farm. Prune the rows of beefsteak tomatoes to fight the verticillium wilt that was making the green forms stunted in comparison to their compatriots on either side, he told them. Set up the small tractor to run a pump to spray neem oil on cucumbers afflicted with cucumber beetles, which lit up like sparks in the sunlight.

Production and water efficiency were a high priority. All the fields were set with drip irrigation line, which conserves vast quantities of water and targets the moisture directly to plant roots. There were lines of San Marzano tomatoes with small mango trees planted among them. Volunteer basil plants occasionally stroked our ankles, permitted to grow wherever a stray seed had taken root. Fields of peppers contained forty-six different varieties, and he filled my cupped hands with samples of Thai black bird's-eye and Trinidadian bishop's crown, the colours and shapes a screaming reminder of the biodiversity lacking in an Indian cuisine that depends on one variety of green chilli capsicum only, one variety of tomato, one variety of onion. There is huge variation in cuisine across India, from the Mughal-influenced meat dishes of the north to the coconut-infused flavours of the southern coasts, but the variety of ingredients is limited, captured in the handful of green chillies tossed in your satchel after buying vegetables from any street vendor.

To jump-start the expansion at FirstAgro, a 6,500-square-foot packing centre was under construction. It is another step towards self-sufficiency in a country that lacks a cold chain. Without this refrigerated supply system from field to market, a large percentage of food harvested in India is lost because of the inability to store or transport it properly.[80] FirstAgro's goal is 5 per cent waste.

Nameet is hugely frustrated by the lack of support for organic agriculture from the government, especially the Bengalaru-based Indian Institute of Horticultural Research.

'The extension office has been closed for a year and a half,' he complained. 'We went there when we started, and they just shooed us away. No one wants to do it,' he said of organics. 'They're afraid.'

Whether it's because they can't get outside funding, or they don't want it, Nameet is thankful to not be accountable to outside investors at the farm. 'I have an ecological responsibility,' he said, 'but no economic responsibility.' He and his brother alone decide how the farm will operate.

But he is frustrated at where the farm subsidies go in India. Why are there no vegetable subsidies? This is a complaint of vegetable farmers in the United States as well, where farm policy leans heavily towards supporting commodity crops such as soybeans, wheat, cotton and corn, the bulk of which are grown to feed livestock or make corn syrup. Meanwhile, fruits and vegetables, the healthy part of the food pyramid, are waylaid, labelled 'speciality crops'. There are some subsidies for horticulturalists in India—FirstAgro has benefitted from some for greenhouses and drip irrigation—but according to Nameet the system is flawed.

We returned to the shade of the main building and sat on the floor, wrestling with the adorable puppies and the question of India's organic future. I asked Nameet if India would starve if it transitioned to organic.

'Yes!' he said without hesitating. 'We just can't produce the yields needed.' It was the complete opposite of what Vinod Jyani in Punjab had answered. Nameet, though a cheerful enough fellow, was pessimistic about the possibility of organic reaching the masses of India's hungry. He thought India was heading for a disaster, 'maybe not in my lifetime, but in my grandchildren's, if I have them. There will be two billion people in India by then. Where will people live, let alone grow their food? It will all be imported.'

FirstAgro was not targeting India's mostly poor majority. Instead, it aimed at a small but growing niche market of elites who could afford certified organic food. But sometimes that's how paradigm shifts begin. It began like that in the West where organic sales have risen steadily since 2002, when the US Department of Agriculture (USDA) established national standards for organic production and processing.[81] Now, organic products can often be found at the tiniest corner store, and prices, which remain higher than conventional, are falling. In the United Kingdom, the organic food market ballooned, from eight million pounds in 1985 to eighty million pounds just five years later.[82] But part of that evolution depended on verifiable sourcing. Consumers have proven willing to pay more—FirstAgro generally prices its crops 20–25 per cent above the market rate of conventionally grown produce—but they also want to know what they are getting. Yet certification in India, Nameet said, is still 'a joke. There are no standards domestically. All exports have to be Codex,' referring to the Codex Alimentarius—meaning 'Book of Food' in Latin—international standards for organic foods established in 1999. This organic standard is one of hundreds created by the Codex Alimentarius Commission, which was established in 1963 by the FAO and WHO to protect the health of consumers and ensure fair practices in the global food market. In the decades since, the Commission has created hundreds of food standards that are based on the best available science and recommendations of independent international risk assessment bodies.

There is also a burgeoning Indian organic certification process, administered by the Agricultural and Processed Food Products Export Development Authority (APEDA), the government authority that runs the National Programme for Organic Production, but even without widespread implementation, the Indian consumer is expressing her desire for chemical-free goods. 'Demand is high,' Nameet said. 'I get calls every day.'

Twice a year, an independent third-party lab that serves as the Codex inspector shows up unannounced at the farm and collects twenty-six samples from across the acreage. Nameet showed me a few Codex reports secured in a three-ring binder, pointing to the lists of 163 pesticides that had been tested for by the inspectors. 'All zeros here,' he said, moving his finger down the page. 'That's what makes me proud. We've always passed, fingers crossed.'

Unlike the farmers I had met in Punjab, who saw their move to organic as a social good akin to fighting for political independence, Nameet was in it for the money and the good food. An unabashed capitalist, he didn't embrace the adoption of pre-Independence modes of agriculture. Nameet shunned even the idea of using jeevamruta. 'It's filled with unknown bacteria and fungi,' he said, 'a completely non-standardized process.'

And he didn't rail much about corporate agribusiness. He wanted little to do with organic activism. He was reticent when I asked him about Vandana Shiva but said she got one thing right: 'Control your seeds,' he said, 'and you control your future.' Even with the lack of diversity apparent at Indian produce markets, heirloom seeds, diverse and indigenous, still survived across India. As for genetically modified varieties, they didn't even warrant his attention.

'I don't believe in that shit,' he said. 'Not because of health, but that they own it. I just ate that tomato. It's in my belly. You're going to say that it's your property?'

What Nameet cared about was making a good living and bringing food to people who cared about it.

'Don't wear the hippie hat, the communist hat,' he said. 'Don't grow food to feel good. Self-help groups are fine, but make it a profitable business that feeds people.'

THE CHICKEN FROM HOME

Far to the north, in the Doon Valley of Uttarakhand, a state tucked against Nepal and China, I stepped into a room whose walls, floors and ceiling were formed from earth and dung. Bouquets of preserved plants hung upside down and the air was dry and fragrant with the residue of herbal life. Patterns swirled on the floor, with Rajasthani designs on the wall. Seeds were held captive in glass jars, in metal canisters, in gourds that rattled when shaken. A prominent board in one corner recorded the number of seed varieties for the last four years: 630 varieties of rice paddy, 126 of rajma or kidney bean, 190 of wheat, thirty-five of basmati rice.

I was standing in the seed bank of Navdanya, a non-profit organization Vandana Shiva established in 1984 to promote organics and biodiversity conservation, support Indian farmers, and rescue threatened crop varieties. Navdanya means nine crops (and seems to closely resemble its founder's name). By 2013, Navdanya had set up 120 seed banks across India, providing seeds to thousands of farmers. The organization conducts experiments on its farm, provides technical support and training to farmers, and even buys produce at a 10 per cent premium, picking it up directly from farmers' fields and selling it at retail shops in major cities.

A young Canadian volunteer named Erin, with straight straw-coloured hair and bare feet, led me and some other visitors across the 47 acres that make up the Navdanya centre. In addition to the seed bank, there were vermicompost bins like shallow graves, covered in burlap and packed with red wriggler worms that turned field waste into prime soil. There was a 1-hectare model farm that demonstrated how to provide for a year to an entire family of six to eight. In one experimental rice field filled with 10 sq. ft plots, seven men and women sliced stems with hand sickles as a man trailed behind them bundling the sample rice stalks and placing them on the metal sign labelling each plot—bhagni naj, naiya naj, jhumri naj. Another man followed with a clipboard, labelling the bundles and making notations. In a short time, the field was reduced to stubble, the ever-present cattle egrets trailing the workers, snatching up bugs stirred loose in the disruption. It was the end of the day and the labourers slipped on their sweaters as the air resonated with the sounds of an evening prayer emanating from a nearby mosque.

A cluster of buildings included a dorm littered with hanging laundry, where Erin and other volunteers, mostly European, lived while they played out experiments with permaculture, the 'permanent agriculture' that draws on organic agriculture, agroforestry, sustainable development and environmental design. A library was full of a hotch-potch of books, from Fritjof Capra to *Hinduism and Ecology*, and a *New York Times* article titled 'Organic Farming Finds Growing Fan Base in India' featuring Navdanya was pinned noticeably on a bulletin board in the main office.

The next day I headed to the climate change office, nestled within a mango orchard set apart from the rest of the buildings, to meet Navdanya director Vinod K. Bhatt.[83] I found him weary at his desk after a rigorous round of travel to Jharkhand and Uttar Pradesh, two of the seventeen states, along with the national capital of Delhi, where Navdanya was active in policy work. He'd been talking to government officials, university professors and farmers about agricultural reform in India. Unlike the young Navdanya volunteers who came and went in waves, Vinod had been with the organization for seventeen years.

'Governments are confused,' Vinod began. His chin was smoothly shaved, his black-and-grey hair mirroring the contrasting navy blue and white stripes of his shirt. Occasionally, a tiny dimple appeared above the crook of his right cheek, above the usual dimple spot. According to Vinod, the governments want to work with multinationals, *and* they want to pursue organic agriculture.

'They want to ride in two boats at the same time. Maybe if they have a good balance, they can do it, but otherwise it is not easy.'

In 2010, Kerala declared that the entire state would go organic. They asked Vandana Shiva and Navdanya to help them make the transition. Sikkim was following suit. Meanwhile Navdanya was also helping Chhattisgarh, Madhya Pradesh and Jharkhand with their organic missions. But efforts were stagnant elsewhere, and the accomplishments were dubious even in those eager states. Vandana had met with the agriculture minister in Bihar as well as Chief Minister Nitish Kumar, but the state's most recent five-year plan made no mention of organic, instead focusing on public–private partnerships, which meant, Vinod said, continuing to depend on the multinationals that had brought the Green Revolution to India in the first place and were now eager to find markets beyond the tapped out landscape of Punjab.

Though state organic movements were stirring, Vinod said, on the national level progress was abysmal.

'We have an organic promotion programme, the National Centre for Organic Farming, but it's like a tortoise,' he said, echoing Nameet.

But he sided with the hopefuls when it came to the question

of transitioning to organic. In fact, he wanted India's prime minister to make the country organic immediately. 'No one will starve. There will be no famines,' he said. Fertilizers cost three times what they did just a few years back. 'Everything is subsidized. If they stopped the subsidies for chemical farming, everything would be organic overnight. If you are addicted to something, then you buy it.'

I thought of Cuba, an entire country that was forced off petrochemical agriculture when the Soviet Union collapsed in 1991, ending a flush era of subsidized fossil fuels and abundant chemical pesticides and fertilizers that the USSR had supplied the Caribbean nation. Cuba basically became organic overnight. But Cuba is a tiny island nation with a fraction of the population that India has, and Cuba, still dependent on imports, struggles fiercely to feed her people.

'The farmers say, "We're planting it because we are getting a subsidy,"' Vinod said of the chemical crops. 'There are rice varieties in Bihar, traditional varieties that produce the same yields as conventional, but because of the subsidies, they lean towards the hybrids. They think if its local, then it's not really good. We've been told for centuries that you are stupid, you have nothing, you are backward, and what you have is really poor,' the local crops dismissed along with the people.

'Now we call them garibon ka anaj—the poor man's grain. Our grain is called mota anaj,' he said, referring to the coarser (and healthier) grains such as jowar, ragi, bajra. 'Whatever is whiter is good, and whatever is darker is bad. Even in grains. White rice is good, and red rice is bad.' Meanwhile, artisanal and heirloom varieties are all the rage in elite Western markets where brown is the new white.

'Under British rule, for a long time, we were taught that whatever comes from the outside is better,' he said. 'Ghar ki murgi dal barabar—the chicken from home is the same as dal. You don't appreciate things from your own home. And the condition of the seed ...' His voice trailed off as he made a what-a-pity sound out of the side of his pursed lip, the dimple appearing. 'It's really sad. Farmers have lost their entire crop. In Bihar, I saw it. They use hybrid seeds, which are designed to ripen in 145 days. But that is only if everything is just right. If something happens in between, drought or too much rain, then these crops fail 100 per cent. Traditional crops have flexibility. Some will delay, or ripen early.'

He dismissed what Amarjeet Sharma and others had said about labour. 'Organic is only labour-intensive in the first year or two,' he insisted. 'There's a maximum loss of 20 per cent in that first year. Only if you're a very ignorant farmer using nitrogen, urea and DAP. But organic costs nothing. You use techniques like jeevamruta and vermicomposting and good manure, green manure. You use the right seed. You'll get 300 quintals of so-called conventional, and maybe 250 for organic,' he said, almost dismissively. 'You save money and time, spend less on health care.'

Vinod was sure that the grass-roots efforts of Navdanya would continue to spread.

'A few years back, in Himachal Pradesh, I met two ladies in Kangra and one of them recognized me from one of our trainings. She said, "Bhai-ji, you trained us. Now we have 2,500 women farmers." Women are taking leadership in Himachal, saying, "We won't listen to our husbands if they resist." Navdanya hadn't given them any support in those two years. They'd learned enough.' Vinod leaned back, a reluctant smile slipping across his face. 'Such things happen.'

I thought of the workers harvesting the test plots of rice and asked if Navdanya had been publishing its results. I wanted to see hard science to balance out the conflicting anecdotes about yields and costs. But in spite of the meticulous monitoring and labels and clipboards and tally sheets, Navdanya seemed to be lacking one critical component that would have made its findings verifiable: peer-reviewed studies to substantiate what they claimed to be finding. Vinod told me that even with 120 staff they didn't have the time to process all the data they collected. They just collected it. Navdanya focused on people and plants, not hard science. They couldn't do it all. It was a labour shortage of a different kind.

'Two types of people come to us,' Vinod told me. 'The first are concerned about the environment and their health. They switch to organic then and there. Others were using chemicals, but it was like gambling. They feel that to switch is risky even if they're losing already. So we tell those people, start small, on just a part of your land, so that you can get convinced. If you're discouraged, then you'll never try it again.' It was smart advice, and resonated with what I'd seen in Punjab, where so many of the KVM farmers had transitioned just part of their land to organic as they tested the impact.

'Farmers won't believe it in books or a lecture,' Vinod said, regarding the lack of science. 'They believe it when they see it. We ask farmers to bring their soil here and then we look at their soil and Navdanya soil under a microscope. They can see that there are more micro-organisms in the organic soil.'

He used the word soil. Not dirt. Dirt is what remains when all life has been sucked out of soil, which is naturally teeming with microscopic beings. Synthetic fertilizers provide nutrients for plants but sterilize the soil they grow within. Those nitrogen-fixing plants such as legumes, along with manure, compost and other organic matter, provide nutrients to roots while also building the fundamental structure of the soil,[84] making it more fertile over time. They cultivate all the teeming life spotted under the Navdanya microscope. Better soil also holds water longer, making it less vulnerable to drought. It's what former USDA official F.H. King saw when he went to China at the turn of the twentieth century, seeking to find out how peasant farmers had maintained soil fertility after working the same fields for centuries. A large part of the answer was to covet manure—human and animal. It was the fuel for permanent agriculture. And the proof was there, alive, under the lens that magnified the magic that happens underfoot when given half a chance.

AGRIBUSINESS TO AGROECOLOGY

So was the Green Revolution a multinational conspiracy to drive up corporate profits, selling farmers on poisons that were killing mothers and their babies and desecrating the lands they lived on? Or was it merely the collateral cost of progress? The famines that wiped out thousands or even millions in one brutal swoop no longer plague India. Punjab remains her breadbasket. Did Monsanto and the rest of the Big Six prevent an untold number of deaths by starvation over the last few decades?

M.S. Swaminathan, considered the father of India's Green Revolution for his instrumental role in bringing the first high-yield wheat varieties to South Asia, provides a cautious answer. As early as 1968—the same year that Indira Gandhi released a commemorative stamp titled 'The Wheat Revolution'—Swaminathan was warning farmers about the limitations of agrotechnology as a solution to world hunger. Beware the transformation of the 'green revolution to the greed revolution' he warned, of adding more and more inputs in the vain hope of eternally pushing the yields ever higher.

'I pleaded,' he wrote in 2010, 'for converting the green revolution into an evergreen revolution by mainstreaming the principles of ecology in technology development and dissemination.'[85] While his advocacy of genetically modified crops remains controversial, his decades-old warnings—about the unsustainable exploitation of land and water, and the vulnerability of monocultures, and the overuse of the fertilizers and pesticides that were so powerful at first—have all proven prescient.

Yet hunger remains. In spite of the high-yield and genetically modified seeds, in spite of the pesticides and fertilizers, a billion people around the world still go to bed hungry.

In India's rural outposts and city slums, I see fewer of the distended bellies that were common on the trips of my youth thirty years earlier, but India continues to have more undernourished people than any other nation on earth.[86] Hunger and the pervasive scourge of malnourishment leaves nearly half of all Indian children stunted.[87] In India, it took looking back at family photos to realize how my 5′ 7″ frame towered over my aunts and uncles. The persistent hunger was visible in the waifish body of the cycle rickshaw-wallah in Bihar, a bright green cloth wrapped around his head, peach fuzz on his lip, who was so light he nearly floated up over his seat as he pedalled. It manifested in the rusty hue of street children's hair. It displayed itself in the slender wrist of a mother as she chopped a fistful of vegetables to feed her entire family in Maharashtra.

Girls and women are often the most vulnerable. More than 90 per cent of adolescent girls in India are anaemic, a direct indication that they lack sufficient nutrition. Diane Coffey of Princeton University found that 42 per cent of Indian mothers are more than twice as likely as their sub-Saharan African counterpart to be underweight, with the average woman in India weighing less at the end of her pregnancy than the average African women at the beginning, a staggering finding.[88]

Lack of food is only part of the cause of this endemic hunger. New science is showing a strong link between malnutrition and poor sanitation. India has the highest bacterial infection rate in the world, due to rampant outdoor defecation and lack of access to treated water. The excessive use of antibiotics has additionally created strains of superbugs that are killing newborns, as well as others. Indians, especially the young, are desperate for a clean environment that doesn't inundate their vulnerable digestive systems with bacteria that make them unable to absorb nutrients, no matter how much they eat.[89]

But 'the food is there', reports the FAO.[90] 'World agriculture produces 17 per cent more calories per person today than it did thirty years ago, despite a 70 per cent population increase.' The problem lies with distribution, politics and corruption, which together create a multi-headed beast. For the farmer, corruption can range from water and land theft to extortionist rates of credit (or no credit at all) to market price machinations.[91] It happens at every level, from local moneylenders up to governments and international aid organizations. Poverty too plays a major role in the inability of people to get the food they need to live. If you didn't grow it and you can't afford to buy it, you don't eat.

But to return to the question that dogged me on my travels in India: Could the world feed itself if it switched to organics only? Is India, or any nation, eternally locked into a system of petrochemical agriculture or is something more sustainable viable? To balance out what I'd experienced from Karnataka to Punjab, I sought out the science I wasn't finding on the ground in India.

The Kellogg Biological Station near Kalamazoo, Michigan, is one of twenty-six global Long-Term Ecological Research sites set up by the National Science Foundation to study ecosystems over decades.[92] Researchers there have been meticulously comparing organic to conventional farming practices, measuring all inputs and outputs, for twenty-five years so far. Gauging rainfall and fertilizer against nitrous oxide emissions, water retention and harvest yields, they found that organic fields produced 20 per cent less grain than those from conventionally grown fields. But they also found that a full half of the nitrogen put on conventional crops leached out—the key source of the nitrate pollution that can cause blue baby syndrome, aquatic dead zones, and a host of other problems for humans and the environment. Kellogg researchers also discovered that in fields treated with smaller amounts of fertilizer and planted with winter cover crops, as is often done in organic rotational planting, the yields were the same as in those of conventional farming practices, but leaching of nitrates was minimal.[93] The ability of the soil to store carbon increased as well, providing a valuable ecosystem service. Here was a happy middle ground.

Meta-studies done in 2012 also determined that organic fields produced 20–25 per cent less than conventionally grown crops. But other studies have shown that organic agriculture requires up to a third less energy than conventional farming systems and that manure and organic inputs stabilized soil and prevented erosion. The science also shows that organic agriculture increases soil microfauna populations and microbial biomass while supporting higher species such as birds as well, as long as the organically cultivated farms were large enough, or there were enough small farms next to one another.[94] But more labour was needed, sometimes substantially more.

Catherine Badgley, an evolutionary biologist working with the Sustainable Food Systems Initiative of the University of Michigan, asked the same question that drove me: Could the world feed itself if it switched to organic methods? In 2007, she published a peer-reviewed paper stating that the answer was yes. The FAO estimates that an average human needs 2,100 calories per day to maintain health. Organic agriculture can produce those calories for everyone. Yes, it would require more labour, but with global unemployment rates on the rise, Badgley suggested, this could be a good thing.[95]

A key caveat to Badgley's findings was that food grown had to be available exclusively for human consumption. The farmers of the world have actually been producing, for a long time, enough food for everyone to eat a hearty, healthy 2,700 calories per day. That is enough to feed nine billion, which is what the United Nations is estimating the human population will reach by 2050. But a third of those calories are currently consumed by cows and other livestock. As the world becomes more carnivorous, this figure will go up. This escalation in meat eating is also contributing to climate change, since raising cows for beef generates thirteen times the greenhouse gas emissions as growing vegetable proteins such as lentils.[96]

Another 5 per cent of agricultural production isn't consumed by either humans or other animals. It is turned into biofuel.[97] In certain breadbasket nations, such as the United States, this percentage of potential food crops not eaten by humans is even higher; in 2012, more than half of US farm yields went to producing ethanol and animal feed.[98] So if counting humans fed directly with products from the field becomes the benchmark of a land's productivity, then India and China actually feed more people per acre of cultivated farmland than the agricultural powerhouse of the United States.[99]

Finally, a whopping third of the world's agricultural yield is lost to waste, the oft-neglected twin in the conservation conversation. The point bears heeding, and not just in the realm of agriculture. Inefficiency—whether in collecting a crop, powering a light bulb, lighting a cooking stove, or running a vehicle—means that you need more of whatever is needed at the source, and everywhere there is waste. Indians lose 40 per cent of their agricultural yields, often because of the missing cold chain,[100] but Americans waste nearly the same amount of their food, even with an impressive string of coolers from field to supermarket shelf. Food worth $165 billion is tossed in the United States every year for imperfections and blemishes, spillage and spoilage.[101] In economically challenged Spain, a cooperative called Ugly Fruit started selling perfectly edible fruits and vegetables deemed too unattractive for the market. Within months after launching in 2013, Ugly Fruit had a waiting list a thousand long.[102]

At the 2007 International Conference on Organic Agriculture and Food Security, the FAO expressed support for a global push towards organic agriculture.[103] The Big Six of agribusiness have little to gain from such a conversion. Their research and development focuses on the microlevel of genetic modification and chemical development. For rich nations, these multinationals pursue development of precision technologies[104] that translate into humanless cabs of colossal combines run by computers, global positioning systems and real-time kinematics that cultivate thousand-acre farms, applying chemicals to pinpoint locations in response to on-the-spot soil testing.[105] For developing nations, they create new products and seeds to entice farmers still recovering from past product failures.

This microlevel research has resulted in impressive gains—crops more resilient to climate fluctuations and better adapted to existing water resources—but the approach is financially intensive and the results are mixed. That secondary pests and uncontrollable superweeds arise so quickly is a foreboding warning.[106] Agribusiness is in an endless race, and evolution is a wily opponent. 'To have risked so much in our efforts to mold nature to our satisfaction and yet to have failed in achieving our goal,' wrote Rachel Carson, 'would indeed be the final irony.'[107]

India, like the rest of the world, does need to extract more from the arable land that remains. Farmland is disappearing as populations grow, pushing the boundaries of villages turning into towns, towns into cities, cities into megalopolises. It's what turned Kushal Pal Singh of Delhi Land and Finance into India's fourth richest man.[108] He said he 'wore kurtas, sat on charpoys, drank fly-infested milk from dirty glasses' in order to win the graces of farming families in Gurgaon. It worked. All that farmland that used to lie south-west of Delhi has now transformed into sprawl, populated by inhabitants in need of more food.

The science also showed that the farmers I met from north to south are all headed in the right direction to incite an organic revolution. While agribusiness technicians tinker in the lab, the Indian farmers are working as agroecologists who aim to create an ecological system in the pursuit of sustainable and economically viable agriculture.[109] Permaculture, conservation agriculture, organic farming and natural farming are all variations of agroecology, which focuses on building soil health by minimizing tilling and growing nitrogen-fixing plants; intercropping compatible plants together and rotating others; using integrated pest management (IPM) and natural biological controls to handle pests and turning to chemicals only as a last resort; improving crop species; and incorporating diversity through tree planting and animal husbandry.

None of this is new. Sir Albert Howard, a Cambridge-educated son of a Shropshire farmer, was the British advisor for agriculture in India from 1905 to 1924, but he realized that India's farmers had more to teach him than he had to teach them. The fundamental underpinning of agriculture was not what grew from the earth, but the earth itself. 'Can mankind regulate its affairs so that its chief possession—the fertility of the soil—is preserved?' he wrote in *An Agricultural Testament.* 'On the answer to this question the future of civilization depends.'[110] To understand a farm, he wrote, it was best to study the forest, an ecosystem where plants coexist with animals and insects. That system manures itself, eternally replenishing its soil with organic matter. Though written decades before the peak of the Green Revolution, Howard's writings remain seminal teachings on organic farming.

Conservation, or no-tillage, agriculture is based on the premise that, according to the International Assessment of Agricultural Science and Technology for Development, it is best to leave soil untilled by maintaining vegetative cover crops and rotating crops in a certain order, planting those that feed the soil before those that deplete it. No-till helps prevent soil erosion, water run-off and flooding, allowing aquifers a better chance to recharge. Across the Indo-Gangetic Plain, conservation agriculture is catching on, with nearly a million hectares of farmland planted using the method as early as 2004.[111]

Such shifts could be crucial to combatting climate change. Scientists are discovering that soil cultivated without synthetic inorganic chemicals is a carbon storehouse of epic proportions.[112] As carbon levels rise precipitously in our atmosphere, causing warming and weather pattern disruptions, the more carbon sinks (storehouses) we have on the planet the better. And soil turns out to be one the best places to sink it. There is more carbon in our earth than in the air overhead or the flora and fauna around us combined.[113] There are 2,500 billion tonnes of carbon in soil, compared with 800 billion tonnes in the atmosphere and 560 billion tonnes in plant and animal life.[114] The Kellogg Biological Station studied carbon storage during its long-term field studies and likewise found that the uppermost portion of earth stores more carbon when it is chemical-free.[115] One Kellogg researcher, G. Philip Robertson, has created a program that would pay farmers to apply less nitrogen fertilizer in a way that doesn't jeopardize yields,[116] rewarding them with carbon credits for good tilth.

Farmers who minimize their carbon impact are working in their own best interest. A University of Leeds study in 2014 showed that even a 2°C rise in temperature will be detrimental to rice, corn and wheat crop yields in temperate and tropical regions. After 2030, the changes will be even worse, with decreases of 25 per cent increasingly common and 50 per cent possible.[117]

The most exciting experiments under way in the zero latitudes are on System of Crop Intensification (SCI), more frequently referred to as System of Rice Intensification (SRI), since the first experiments were done on paddy.[118] Originally developed in Madagascar thirty years ago, the concept is basic: instead of broadly sowing seeds and hoping for the best, farmers carefully cultivate seeds until the plants are established and then transplant them out into the pockets of rich soil in the fields to mature, applying targeted water application as they grow.[119] It's a quality-over-quantity approach: plant fewer rice seedlings, and tend them better. The method has shown success with not only rice, but wheat, sugar cane, ragi and other crops around the world.[120]

According to Cornell University's SRI International Network and Resources Center, this method has been adopted by at least 9.5 million farmers in India, Cambodia, China, Indonesia, Vietnam and other rice-growing countries.[121] A recent World Bank study in Bihar found huge productivity with the method, with rice production rising 86 per cent and wheat production 72 per cent.[122] Likewise, a large-scale study commissioned by the UK government's Foresight: Global Food and Farming Futures project demonstrated that a move towards agroecology more than doubled crop yields over a 3–10 year period on thirty-five million acres in Africa.[123]

Those figures are game changers.

There are hopeful reports of SRI popping up across South Asia. In 2013, food policy analyst Devinder Sharma asked Chief Minister Nitish Kumar of Bihar how he intended to ensure that his state did not end up depleting its groundwater like Punjab. 'I am aggressively promoting System of Rice Intensification (SRI) method of cultivation,' the chief minister replied, 'which reduces water consumption by 50 to 60 per cent.'[124] The latest Bihar government plan may have not included mention of organic agriculture, but there was a cash incentive of 1,200 rupees along with improved seed offered to farmers adopting the SRI water-saving technology.[125] By 2013, over a hundred thousand small farmers across nine districts in Bihar had adopted this new system of seed treatment and planting that uses no chemical fertilizers or herbicides and plans were in place for SRI expansion in 2015 across all of the state's twenty-eight districts.[126] The Bihar government claims that hundreds of thousands of hectares are now planted using SRI methods, with average yields at least 40 per cent greater than for conventional rice farming.[127]

In Gujarat too paddy cultivators in the Navsari and Dangs districts are working with the Aga Khan Rural Support Programme to spread SRI methods and are claiming 40–150 per cent increases over conventional methods.[128]

In Tamil Nadu, a rice paddy grower apparently set a record. S. Sethumadhavan from Alanganallur claims to have harvested a record yield of nearly 24 tonnes of paddy rice per hectare using SRI method, though a few have said his story sounds too good to be true.[129]

In a proactive measure in Punjab, the Punjab State Farmers Commission released a draft of a new agricultural policy in 2013 that would attempt to ease the drain on water resources by diversifying crops and reducing the acreage under water-intensive wheat and rice paddy production.[130] The 2009 Punjab Preservation of Subsoil Water Act forbids rice transplantation before 10 June in an effort to grow this thirsty crop during monsoon, when the most water is available. Nascent efforts at SRI, along with direct seed rice (DSR) cultivation, which involves planting of seeds in rows at a specified distance, are also arising. Governments are providing subsidies to farmers who adopt DSR with a condition of no straw burning. One DSR farmer in Moga district said the method saved him Rs 16,500 per acre,[131] which is a substantial sum for farm families that might only earn an average of about Rs 75,000 per year, at best.[132]

Worldwide, thirty-seven million hectares, less than 1 per cent of global agricultural area, are being worked organically.[133] But that figure is triple what it was at the eve of the new millennium.[134] Most of those lands are, so far, in the developed world. For the six hundred million farmers in India, it is a moment of possibility.

The maze of acronyms—SRI, IPM, DSR—represent a spectrum of tactics that can be used to move towards sustainable agriculture. If implemented on a broad scale in India, these methods could produce enough food for a growing nation while lessening toxic exposure for life forms from honeybees to humans, preventing the poisoning of water and land, alleviating farmer debt, easing climate change, and cultivating food free of chemical residues. Each of these outcomes is itself monumental. Collectively, they could transform the way we humans feed ourselves on the planet.

But that is the big picture. Closer to home and hearth, at the small scale where real change is most likely to happen in India, these methods might allow KVM to find companions along its agricultural road. In a world in which India's farmers turned to organic with as much enthusiasm as they embraced the Green Revolution, Gora Singh, with the help of his spiders, could continue to honour his dead mother. Amarjeet's sorrow might slacken. Ashish could enjoy his coffee knowing his workers no longer think he's crazy, and Nameet could relish his glass of shiraz as FirstAgro's Karnataka farm—and others like it—flourished. A conversion to organic could also mean fewer empty containers of monocrotophos lying in wait to be filled up with cooking oil destined for a schoolchild's midday meal in Bihar or anywhere else.

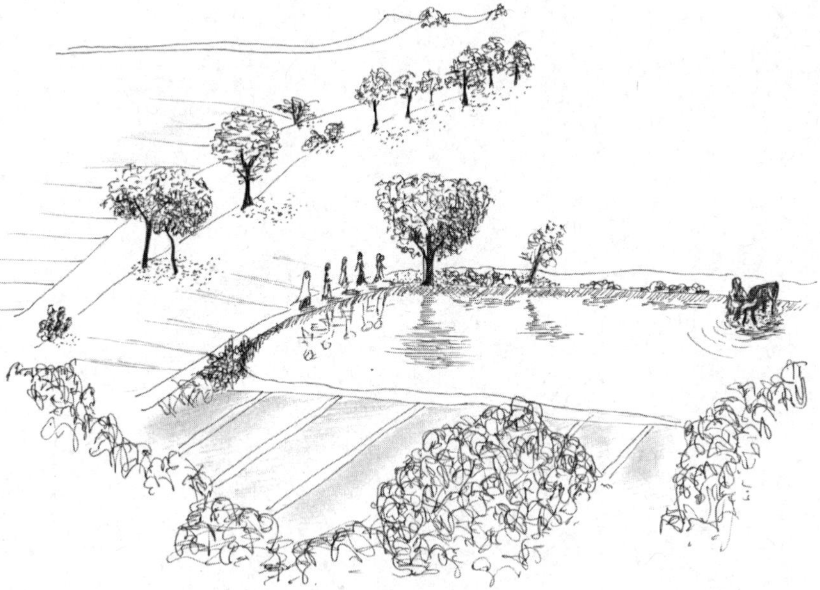

2
AP
WATER

We are unable to recount the true story of who had it dug, or when.
It is said that demons, the Asuras, dug it themselves.
That could well be true. Could humans like us
dig such an immense pond?[1]

—Fakir Mohan Senapati

Thousands have lived without love, not one without water.[2]

—W.H. Auden

A RIVER RUNS AGAIN

Bhaonta is nestled within a river basin that disappeared and then returned in the Alwar district of Rajasthan, located in northern India, just a few hours south-east of New Delhi. Visitors, for the most part, bypass the region. They might stop through Sariska Tiger Reserve, known for its occasional lack of tigers, in hopes of spotting the stripes amid the dry deciduous forests of dhok trees and thorny undergrowth. But mostly, the hordes flock predictably to the south-east, to step inside the marble mausoleum of the Taj Mahal, or south-west to Jaipur's pink palaces.

I went to Alwar in search of water, lured by the accounts of the Rainman of Rajasthan that appeared in the papers, recounting tales of a man and his non-profit organization, and a desert landscape reborn. More often the news about water in India was less hopeful, stories of waterbodies under assault, dying from overuse and exhaustion, from pollution too often disguised as devotion. Repeated efforts to restore the holy and wholly polluted waters of the Ganga from an onslaught of industrial, residential and sacrificial effluents were repeatedly failing. The effect of untreated sewage discharged along the 14-mile stretch of the River Yamuna that runs through New Delhi prompted the United Nations to declare the river downstream dead.[3] Developers routinely destroy wetlands as they erect buildings that invariably get flooded, and the marshes that still remain are often choked with human refuse of every form. The stories I heard in Punjab of dried-up wells were repeated in Rajasthan, where the western side of the state was so dry that it transforms into the pure golden sand dunes of the Thar Desert. The Alwar district in the north-east at least had rain.

But India's largest state of Rajasthan has always been vulnerable to droughts. At the worst of times, people and governments clasp locks on pump handles. Villages might only receive water every fourth day, via tanker trucks that are circled by thirsty people, elbowing their neighbours to get to the front of the line. There have been reports of caste and ethnic communities aligning themselves, capturing water tankers and keeping others from accessing them, and stories of families who sold their jewellery so that they had money to buy water. Camels that used to transport people are used to haul water tankers instead.[4]

I'd heard things were different in Alwar, and on a late fall morning, I headed to the village of Bhaonta with Kanhaiya Lal Gurjar, a lifelong resident and the general secretary of Tarun Bharat Sangh (TBS), the non-profit organization that Rajendra 'The Rain-man of Rajasthan' Singh had formed in nearby Bheekampura. For thirty years, TBS had been coaxing the local Arvari river back to life in the dehydrated landscape, along with the other rivers of the region—the Bhagani-Teldehe and Sarsa, the Jhajwali and Ruparel—all tributaries of the great river Yamuna that flows towards the Ganga.

Kanhaiya had been working with TBS since nearly the beginning, and Bhaonta was his home. We travelled down a one-and-a-half-lane road to get there, between fields of cotton, maize and vegetables. Jeeps passed laden with people, bodies bursting from the seats, perched on the hood, and casually sitting upon the roof with knees jutting skyward. Periodically, we wound through lanes between village homes, a place not built for cars, buses or trucks, though all slipped through, barely. Air horns ricocheted off rows of buildings with concrete walls and straw-thatched roofs. Half a dozen men sat in circles playing cards. An old woman smoked tobacco from a pipe as young ones filled water from a well. Men and women labourers circled between a parked truck and a field, slowly filling the truck's bed with bundles of maize straw they carried in sheets of cotton. It was only filled a few feet high early in the morning, but by mid-afternoon, a cloth extension had been added to the truck's back end, which ballooned impossibly with the material heading to market.

After a few turns off the main road, we entered the outskirts of Bhaonta. Some women were busy in the fields while men laid blue irrigation pipe in a shallow ditch. A cell tower loomed from a field. We pulled over at what seemed like a large pond, quiet and still. The hills of the Aravalli Range rose up in every direction around us, enclosing us in a pastoral valley. Fields extended towards the hills, which contained abundant copper-and-pink marble exhumed for travel to distant affluent bathrooms where water was a bounteous resource. On the far shore of the reservoir, a woman in a bright orange sari stood watch over the dark forms of three water buffalo as they quenched their thirst. We stood upon an earthen dam that held the body of water in place.

This is what I had come to see. Indians travel to America to gaze upon the staggering cascades of Niagara Falls and the spectacle of the Glen Canyon Dam, but I had come to this part of India seeking humbler bodies of water, constructed and crafted by simple people in the hopes that they could resuscitate their homelands from pervasive drought. Rajasthan was always a dry place, but the droughts of the last century seemed to suck the life out of the region. There were years when the monsoon failed to produce even one good downpour. Millions of cattle were left to die after prayers for rain went unanswered, their fodder and watering holes vanished. Weddings were postponed indefinitely. Men left in droves for the city, the agricultural lands barren without water.[5]

I came to see the opposite of mega dams and grand interlinking river project schemes. Instead of constructing epic waterworks, the people in the small village of Bhaonta, under the guidance of TBS, had sculpted earth and other materials into dozens of small dams over the course of a generation. These small water-retaining structures go by many names: johads, check dams, jorees, bunds, khadins. Each johad has a name, and Kanhaiya introduced me to the one we stood upon, called the Naharsingh Walla Johad, after a tiger god.

Then he led me to a much smaller johad, a few hundred yards away. This, he said, was the Bandi Johad. It had just a trace of water in it, Kanhaiya explained, because the rest had already been used up for irrigation. A substantial tree—the sacred fig *Ficus religiosa* with large heart-shaped leaves and its distinctive drip tip—stood sentinel, and Kanhaiya walked towards it, curling his fingers around one of the many thick trunks that emerged from the tree's base.

'I planted this tree,' he said, his gaze shifting to the remnants of the water below us. 'This was the first dam I built, hai na?'[6] He ended most of his sentences this way—'isn't it so?'—as though beckoning the listener to understand. Kanhaiya, also the local name for Lord Krishna, was a small man in his late forties, with a delicate and gentle air, the hair on his head mostly salt, his mustache dominated by pepper. He was a serious man, introspective, but at moments an abundant smile would envelop his face and it appeared there as he made contact with the tree, his tree, and a starburst of creases splayed from the edges of his eyes. The small pool of water we looked over, no more than a large puddle, was not particularly striking or beautiful, but the droplets it contained, and the life it supported, changed Kanhaiya's life, and that of his village.

The last prince to rule over the region before the titles of royalty were banished with Independence was Maharaja Tej Singh. Like the princes before him in the area of the Aravalli Range, he was an Anglophile, and had even been knighted, becoming Maharaja Sir Tej Singh. But to Kanhaiya and the people of Bhaonta, he was known as the prince who cut all the trees. Under his reign in the 1930s, the forests were felled. The underground network of roots that had held the precious desert water like a sponge broke down. Above ground, the River Arvari ceased to flow anywhere on its 56-mile length by the 1930s, according to TBS.

It is a bad time for rivers to be vanishing. In Rajasthan, and elsewhere, the world is becoming thirstier, with human need for water on the rise.[7] Globally, the use of freshwater has been growing at more than twice the rate of population increase over the last century.[8] According to the United Nations, 1.8 billion people will be living in regions with absolute water scarcity just a decade from now. They will have less than 18,000 cubic feet of water available to them over the course of the year, less than a fifth of what the average American uses today.[9] In that same ten-year time frame, two-thirds of humanity could be dealing with daily water stress conditions of less than 60,000 cubic feet per year.[10]

But in the life of a poor person, these technical distinctions simply mean they lack access—physically and/or economically—to the most basic of resources: clean water for drinking, cooking and bathing. It means they reach for the water they can, which is often far away, or the muddied dregs shared with too many others, or waters contaminated with disease. Scientists estimate that a third of India's aquifers will face complete depletion in the near future, and a quarter of the country's districts face the prospect of a drought in any given year.[11] The peninsular nation of India is a land bounded by water, and nary a drop to drink; only 2.5 per cent of all water on earth is fresh and not saline, and two-thirds of that is locked up in glaciers and ice. What remains travels through a recirculating system of evaporation and condensation, up into the air above and back down, upon our heads.

In India, rains arrive in the form of monsoons, heavenly replenishment of water. Never all that predictable, they seem to be getting even more erratic, possibly because of climate change. Three quarters of South Asia's rainfall occurs between June and September, most of it in the span of fifty-four days, dangerously concentrated even more within a time frame of about one hundred hours.[12] In 2014, floods devastated Jammu and Kashmir, and parts of Uttarakhand were swept away after three days of monsoon rains in 2013 that were four times greater than the average accumulation.

Climate change is also hastening water scarcity as it impacts the other major source of South Asia's freshwater: glacial run-off from the Himalayas. As the warming of our atmosphere erodes the Himalayan glaciers at high elevations, crops desiccate in the plains and people go thirsty. Meanwhile, along the thousands of miles of India's coastlines, sea level rises.

If the water from above is increasingly undependable, the water underfoot is no less worrisome. The plummeting water table of Punjab resembles the falling water tables in Rajasthan and the North China Plain and Saudi Arabia and California's Central Valley.[13] Water everywhere is increasingly scarce, and agriculture is by far the largest user of the world's freshwater.[14] Withdrawals in many water basins, warned the 2012 World Water Development Report, are exceeding the rate of recharge and are, to put it bluntly, unsustainable.[15] Asia dominates the consumption, accounting for 80 per cent of the total area irrigated by groundwater in the world.[16]

With Independence, India saw a surge of irrigation projects for agriculture as the nation strove to rid its population from the threat of famine, and it was then that the open wells that had provided for humans, animals and plants started to run dry. In the early 1950s, only dug wells were used to extract groundwater for irrigation and domestic use, and almost all were of a shallow depth. Water was lifted manually, with animal power, or the circular lift of graceful troughs on Persian waterwheels.

A decade later, the government became proactive about helping farmers extract more groundwater in the quest for food security. The memory of the millions who perished in the catastrophic Bengal famine of the 1940s was still fresh.[17] Electric pumps made the task easier, and subsidized electricity for farmers fuelled the increase as water was extracted from greater depths.[18] Development funding, both public and private, flowed in, and bore wells tunnelled into the earth, tapping into underground aquifers that had been accumulating for eons. The wells, 8-feet-deep or 80-feet, would run for a handful of years and then go dry. The well diggers dug deeper—120 feet, 200 feet, 400 feet. Advances in technology made the digging easier, but it didn't alter the fact that the water would again be quickly depleted.

Between 1950 and 2007, the amount of cultivated land in India increased by only 10 per cent, yet the percentage of that land that was irrigated increased from 16 per cent to nearly 50 per cent, and the vast majority of the increase came from tube well extraction.[19] With the Green Revolution of the 1960s and

1970s, year-round crops were introduced, as I'd seen in Punjab, along with year-round irrigation, though the rains continued to fall from the sky only during the monsoon months, during those scant one hundred or so hours, such a spectacularly short amount of time to replenish deep aquifers. Recharge from glacial run-off is similarly troubled, as the glaciers shrink in size each year.

India's post-Independence expansion of water access was magnificent and unprecedented, yet this progress is tempered by a growing list of concerns. There are striking gaps between water allocation targets and actual achievements, delays in projects and spiralling costs. There is also the consistent failure to address environmental and human impacts of water projects.[20]

In the Alwar district, the River Arvari had been gone for decades, but by the 1960s, the groundwater was depleted too, and then migration on a major level began. For the men of Bhaonta and elsewhere in Alwar, tilling the fields of their fathers was pointless without water. By the 1960s, you could find the men, instead, by Lahori Gate in New Delhi, where they loitered about, ready and willing day labourers. They hoped for work so they could send money to the wives and children they left behind in a diminished landscape, families fractured. In the quiet villages, those that remained grew what food they could and spent more time each day travelling to distant wells to fill vessels that would burden their heads and shoulders on the trip back home. Life in the village is dependent on the land, and land is dependent on water. Life in the village, never easy, became harder, if not impossible, without water.

Things weren't any better by the late 1980s when Kanhaiya Lal Gurjar was on the brink of adulthood and preparing, like most boys his age who hadn't already left, to head to the city in search of work. Then, on 2 October 1988—Kanhaiya remembers the date—his uncle, a Gandhian and social activist, took him to meet a man named Rajendra Singh of the nearby non-profit Tarun Bharat Sangh. Rajendra had not yet earned his nickname, the 'Rainman of Rajasthan', but the TBS founder had begun building small dams nearby a few years earlier, capturing monsoon rains so that the water could percolate down into the groundwater and not be lost to surface run-off. Stories about dry wells springing back to life were circulating through the Aravalli hills. Rajendra had talked to Kanhaiya and others about the importance of building johads, saving the forest and protecting water.

'I got motivated to do this work, hai na?' Kanhaiya said.

'We are a backward caste,' he added, 'the Gurjar caste,' and he said his prospects as a boy had been slim. He'd been lucky to receive some education and he became a teacher. After hearing Rajendra speak, instead of moving to New Delhi, he joined TBS. He has spent the last twenty-five years helping to revive ancient water management techniques—much of what TBS is doing is not new—in which the seemingly simple act of digging a pond to capture monsoon rains radiates out in a replenishing wave across the landscape. Through these small-scale but broadly implemented water restoration efforts, Kanhaiya and others have transformed villages around the region. Between its founding in 1985 and 2012, TBS, in collaboration with locals in each place they work, has built more than 12,000 johads and other rainwater harvesting structures in thirteen districts across Rajasthan with national and international donor funding. The johads in Bhaonta were some of the first constructed. In 1990, when Kanhaiya was twenty-three years old, the River Arvari, which he'd never seen in his lifetime, began to flow again.

Kanhaiya released his hold on the grown tree he'd planted and reached for a small unripe fig hanging above his head. He turned it inside out, revealing the packed interior, abundant with tiny seeds still forming, all potential. Every completed johad has at least one tree planted on its banks, he explained, so the roots can hold the soil together and the canopy can provide shade to minimize evaporation. Below our feet, the tree's roots were soaking up moisture, reinforcing the cycle of water retention, inching absorbent tendrils through the tough ground. Other trees sprouted nearby—neem, banyan and gular trees, seemingly growing from rocks—reversing the clear-cutting of kings. Birds flitted between branches.

'My father used to see tigers here.' Kanhaiya said, as he led us back towards the first johad, where we'd parked the car. The path between the ponds was dry and rocky though water was so close. Kanhaiya pointed out how the johads aligned: one, two and a third beyond our sight, all in a row, with overflows so that when one filled up, it fed into the next. Before the placement of the johads, there would have been no pockets of moisture, just a continuous swath of rock and dust where monsoon rains could cascade, causing erosion along the way.

Though it was late fall, the sun beat down on us. Kanhaiya, shrouded in a traditional white cotton kurta with a scarf loosely wrapped around his neck and sandals on his feet, seemed unaffected by the heat. He pointed to where his father had seen a tiger, the inspiration for the name of the Naharsingh Wala Johad. As the waters and trees returned, there had been tiger sightings once again after a long absence, the return of water serving more mammals than just the humankind.

Next, Kanhaiya took me to the centre of the village, to the place that had once been desolate and bereft of men of a workable age. I saw males, young and old, in the fields. I heard the sound of a saw penetrating stone, as an artist crafted marble statues, dusting himself and his surroundings with a fine layer of airborne marble, as airy as talcum. Two women dressed in red leaned in close over a baby swathed in grey fabric that one of them cradled in her arms; men and women were most certainly living together in Bhaonta.

In the heart of the village was Kanhaiya's house, a pale pink concrete structure with a teal porch, ceiling fans and lights. His wife, Rama Devi, her head covered with the edge of her red sari, was in the kitchen, sitting on the floor tending an open fire in the chulha, and bands of smoke refracted in the light from the window. Lada, one of their five children, was an eighteen-year-old moon-faced girl with her father's full lips, and she was having a hard time keeping her pink head-covering in place as we chatted for a few minutes.

Kanhaiya wanted to take me to a showcase johad of the village, a thirty-minute hike uphill from his house. He led me behind the house, past a few water buffalo that were standing in their stalls, and then we quickly began to ascend a steep scree-surfaced hillside dotted with scrubby shrubs. After ten minutes we had gained a bird's-eye view of Bhaonta, and we paused to catch our breath and gaze at the valley below.

It appeared as Eden. Harvard professor Diana Eck writes about Anandavana, the forest of bliss, a place of 'groves, streams and pools'.[21] She is describing ancient Benares, today's Varanasi, on the banks of the River Ganga. It was a place far from Kanhaiya's village, but the concept transcends geography. Imagine a land so unspoiled—Jean-Jacques Rousseau's romantic untouched landscape, John Muir's Yosemite—that it created a natural garden so tempting that the gods were willing to leave heaven to inhabit it. The Puranas describe the forest of bliss as a place populated by ancient deities called yakshas who were associated with trees and ponds, and their female counterparts, yakshis, who 'were first depicted as tree-spirits: shapely, full-breasted and full-hipped women, embracing the trunks of flowering trees'. (It was the original Chipko movement, when Indian women, the original tree-huggers, embraced trees with their arms and bodies to prevent men from cutting them, in India's recent and distant past.)

But which came first? The canopied trees and crystal-clear ponds, or the gods and goddesses and sacred places? Even as the number of humans increased and the virgin landscapes vanished, roadside shrines remain tucked along the protruding roots of great trees or the banks of rivers. Stones serve as religious icons, at the heart of a hidden Hindu temple or Al-Masjid al-Haram. Water becomes a blessing, a purifier, a message from God. One bows prostrate and makes intimate contact with the earth.

There are too many humans on earth to hope any longer for great expanses of land or even sea spared from human impact. But can't we treat the places we inhabit with reverence? Looking down upon the bucolic valley of Bhaonta, Kanhaiya and I saw a verdant place, an Eden reborn.

'There has been a topographic transformation of the area,' Kanhaiya said, as we took in the green and golden fields. He told me how most of the farmland below us had become fallow during the decades of drought. How many householders were missing from the houses, whose roofs we looked down upon, that ringed the fields—homes that belonged to Kanhaiya and his neighbours and were partly obscured by the foliage of trees that grew among them. From our vantage, you could see the village lands in a small valley nestled between the Aravalli foothills, the buildings hugging the higher ground and open farmland extending beneath. Lines of trees marked the divide between fields, and Kanhaiya showed me how the water moved. Off to the right, we could see where the trio of johads were, on a subtle uphill. They captured the first ribbons of water that reached towards them. Pointing at a low line of raised earth that ran the small breadth of the valley, bisecting it, Kanhaiya explained, 'The bank was built in 1993, and it's just a few feet high, but that's enough to catch the water when it comes. All the land to the left of it'—lush farmland indistinguishable from what I saw to the right of the berm—'was barren, hai na? We block the water and then drain it right before planting,' he said.

A line of women, a brush stroke of six bright saris—orange, yellow, red, green—moved in single file along a path near the low line of raised earth, a splash of colour passing tiny figures of men in earth tones bent at their pastoral tasks.

Today the villagers cultivated fodder for their livestock, mustard for oil, barley and wheat, and vegetables such as potato, onions, bhindi, cauliflower. Before the johads, Kanhaiya said, they could only grow maize, millet and barley in the kharif season of the rains, but now they could grow wheat or mustard in the dry season of rabi too. The average family in the village had 1 hectare and grew more for subsistence than for cash. Kanhaiya's extended family had 10 hectares, and he cultivated 2 hectares for his family, each generation of sons dividing the land into ever smaller parcels.

Kanhaiya brought my attention back to where we were standing. In a slight depression, an admixture of ochre earth and chipped rocks accumulated along the edge of the steep incline. At the centre it was damp from no discernible source, and the few goats that were grazing there left coffee-bean-shaped imprints of their cloven hooves in the moist earth. This was a joree, Kanhaiya said, a little johad that was built a couple of years earlier and named after his late father. When full with rainwater, it stretched 30 feet across. Saplings had been planted. The fledgling foliage of two trees—one banyan and one neem—was barely visible, contained within the bounds of a mountainous pile of thorny branches to protect them from the hunger of goats and other browsers.

'Before the johads,' I asked Kanhaiya, 'what was it like?'

'We never had a shortage of drinking water,' he said, 'but we didn't have irrigation water and the animals could only drink what we could draw from the well. But after the dams, the wells were full all the year round and the animals could drink from multiple places. Now, they can drink anywhere, and there's enough for irrigation.' He bent down towards the thorny branches, peering through them into the small saplings and then looking out to the valley again.

'Now there's so much water that even if we have a drought, we can handle it for two or three years. When I was young, about 50 per cent of the young people migrated away for work, hai na?' he said. 'Now, maybe 5 to 10 per cent voluntarily migrate away, because good schools and work are in the cities. Some of the kids I taught are doctors and professors now. Labourers will migrate out seasonally, maybe just for a few months, especially if there are extra people in the family.'

'Earlier, they went because it was the only way to earn money,' Kanhaiya said. 'Now they go to earn extra money.' To leave, or to stay, had become a choice.

Though the valley was green, it was dry and rocky as we left the joree and continued uphill to the larger johad he'd told me about. The terrain grew open and seemingly empty except for occasional calico-coloured flocks of goats that added colour to the drab landscape, their coats polka-dotted with fluorescent paint for identification. Some shepherds let their goats wander alone, the threat of jackals or hyenas or leopards a nocturnal worry only, but one shepherd accompanied his large flock, the bells around their necks and their baaaahs a symphony in motion. He stopped one doe and bent to fill his cup with the freshest of milk for a drink.

A cluster of bright colours moved in the distance. As we got closer, I made out thirty women carrying shallow dented metal basins filled with stone and dirt upon their heads. Bhaonta's building of johads was not done. The women's loads were for a johad they were constructing 100 metres away. They moved with the organization of ants, methodically travelling in a line, wielding pickaxes and filling their discs with the excavations. The women were not from this village, I was told, but came from nearby as part of the Mahatma Gandhi National Rural Employment Guarantee Act (MGNREGA), a law passed in 2005 that ensures at least 100 days of guaranteed wage for each household that volunteers for unskilled labour pools. The women were earning 135 rupees per day for their manual labour, supervised by a few men standing with clipboards in hand.

We left them to their work and continued farther uphill. The higher we climbed, the smaller the few trees became, until there were just desert shrubs. The abundance and girth of the stones grew, and rock faces reached vertically up from the sides of the well-trodden villagers' path. The landscape reminded me of the red rock deserts of the American southwest, the sheer walls closing in on us as the path became steeper and narrower. We entered a canyon and then one turn revealed a dam larger than all the others Kanhaiya had shown me.

The Sankara Dam was 70 feet across and 30 feet high, made of large stones mortared together with concrete, buttresses providing extra support. The contained water measured at least 100 metres across, and on the far side, people watered their animals. Dozens of dragonflies zipped through the air, landing along the calm shoreline. Around the water, trees flourished.

I asked Kanhaiya if they'd ever had a johad fail them.

'No,' he answered quickly, 'we know what techniques to use.' He took my notebook from my hand, pulled his blue pen from his shirt pocket, and began sketching the watershed. This, he tapped with his pen point, was the obvious point where the johad should be built, and here is why the johad should be concave, bowed towards the incoming pressure of the water that builds up behind it. And with a few more strokes of the pen, he showed me where three more dams were, farther uphill. Like the trio he'd shown me below, each one fed into the other.

'This area is a public sanctuary,' Kanhaiya said. 'We've decided to not cut any trees. If we need wood, there is always enough available without cutting.' He pointed out neem trees that someone had planted on the communal land.

Back in 1987, when TBS granted the request to come work in Bhaonta, there were stipulations. The villagers had to be involved in a substantial way, contributing either money or labour. For the large dam before us, TBS gave the cement and their guidance, but it was the villagers who did the work. Kanhaiya said that fifty-three village families each contributed over 100 days of labour.

'They have such zeal for it,' Kanhaiya said, 'and somehow, when they don't have other paid work, they come.'

The sky was a brilliant blue. I had been travelling around northern and central India for months and everywhere a haze had hung over my movements. Air pollution was reaching record levels across the nation, and not just in the cities. But in Bhaonta, the smog was gone, and the sky felt scrubbed, in sharp relief against the rocky cliffs that reached into the sky.

Looking overhead, and breathing the clean air, I asked Kanhaiya if he had seen a change in monsoon patterns in his lifetime.

'In 2004, this was overflowing,' he answered. 'Last year, it was only half-filled. But the rainfall has decreased over time. And more than that, it has changed when it comes. Before, we might get 600 millimetres over four months. Now, like this year, it can come in just two to three weeks because of climate change. It came forty-five days late this year, then all at once.'

Two women headed downhill, smiling at us, flip-flops on their feet and lofty bundles of greenery balanced on their heads, bound for hungry livestock. A small raptor the size of a sparrowhawk darted from one side of the canyon wall to the other. Kanhaiya watched me with interest.

'You should see how beautiful it is in the monsoon,' Kanhaiya said, and images from Satyajit Ray's classic film *Pather Panchali* popped into my head, of men, women and children tilting their heads back to let long-awaited waters drench them. For desert dwellers especially, the monsoon is a blessing, quenching the otherwise unquenchable. 'I come up here just to sit and enjoy, hai na?' he said.

We began our descent, back towards the green of the fertile fields in the valley below. Past the village temple, where Kanhaiya paused to lean his forehead against the orange wash of a deity's icon, and past the stone wall that enclosed a sacred grove of trees the village planted. Past a boy taking a midday bath at a well where abundant water gushed out from the pump and continued on its path towards a muddy field where boys and men gauged furrows in the earth with their hands so the water could flow. Peacocks strutted in the background.

The Arvari remains a perennial river that surfaces seasonally at its upper stretches, but the full wells attest to the replenishment. This may have once been the domain of princes, but it was ordinary people who lifted the shovels that reshaped the land's destiny.

WATER SPOUTS AND SERPENT MOUTHS

Volcanoes, landslides, glaciers and beavers did it first. By design or by happenstance, they transformed their surrounding topography and stemmed the flow of rivers and streams, which pooled up behind the blockages, creating lakes, ponds, wetlands and a cascading set of ecological changes. But humans too have been modifying the flow of water around them to suit their needs for millennia, plugging here and siphoning there. The archaeological record has found evidence of complex systems built by humans to harness water as far back as the third and fourth millennium BCE. Human hands have used earth, timber and stone masonry, and later added concrete and steel to construct dams, levees, dykes and weirs. Our objectives in the quest to control water have been diverse: block the flow across land or catch the rains from above, control floods, secure water for crop irrigation, or spin turbines to generate electricity. Some structures are tiny and others are quite long—one in India stretches sixteen miles across—and some are massive enough to be seen from space.[22]

In Mesopotamia's Euphrates–Tigris River basin, early peoples drained marshes, constructed dykes and canals, and built water diversions. In north China, prehistoric civilizations built massive embankments to control the course of the Yellow River, and in South Asia, inhabitants of the Indus River basin constructed elaborate networks of canals.[23] Throughout the Kathmandu region, there was once a valley-wide water management system that arose in the first millennium BCE, the remnants still revealed in the square hitis, sunken water sources enclosed in brick, and the dhunge dharas, stone water spouts with carved serpent mouths.[24] In south India, the Chola King Karikalan diverted water for irrigation from the Cauvery River with the Grand Anicut, a 1,000-feet stone dam built nearly 2,000 years ago. One of the oldest and largest surface systems of its kind, the Grand Anicut is still functional today.[25] Some of the open water tanks of the south, and the catchments (ahar) and channels (pyne) of Bihar, as well as ingenious water-retaining devices used in Gujarat and Rajasthan also remain in use in the twenty-first century—hundreds if not thousands of years after they were constructed.[26]

But to consider the elaborate water systems is to remember how easily knowledge can be lost as well as gained; knowledge is as peripatetic as water. In what is now Pakistan and northern India, there were toilets and sewage systems 3,500 to 4,000 years ago; they were discovered by architect Sudarshan Raj Tiwari when he went sifting through inscriptions, chronicles, legends and rituals to piece together the daily lives of the ancients of the Indus Valley culture that produced the Mohenjo-Daro and Harappa civilizations.[27] Today, a third of Pakistanis and half of those living in India defecate in the open.[28]

Reminders of water systems from times past are also hidden within India's largest cities. In the hip Hauz Khas neighbourhood of New Delhi, one can sip cocktails and listen to live music at a gastro pub that overlooks a pool of water constructed in 1295. Once, it was the source of irrigation water for fields ploughed centuries ago, and was used to cool royal pavilions now in ruins or vanished beneath the foundations of nondescript concrete edifices. Then, it was fed by seasonal streams, but now the pond fills with raw sewage that arrives from a treatment plant a few miles away, and the water's hue is a particularly troubling shade of green.[29]

Author Cheryl Colopy describes New Delhi's Ugrasen ki Baoli, a water cistern hidden behind buildings across from the Fulbright office on Hailey Road. A 'subterranean construction of stone descending five stories lay at my feet', she wrote of the space 60 feet wide and 200 feet long.[30] It is the remnant of an ancient Delhi unrecognizable today. Once, the tank held water in its descending layers of archways and stairs, but now should water find its way there from the concrete above, it is promptly pumped out, lest it flood the nearby Metro tunnels.

In the Vedas and Mughal history and secular literature, there is evidence of royal support for waterworks. Constructing and improving access to water was a benefit a king bestowed upon his subjects.[31] Brilliant waterworks, physically complex with impressive institutional arrangements, once existed here.[32] So what happened? How did all these systems crumble into disuse?

In India, the British had fully usurped power from the Mughals by the nineteenth century. In the early stages of the colonial era, the new administrators actively renovated and improved water systems, and—as technology progressed—brought modern-era large-scale surface irrigation projects to India. In the mid-1800s, the Upper Ganga Canal, the Upper Bari Doab Canal, and the Krishna and the Godavari delta systems were constructed with anicuts, small dams built across rivers to divert flow to elaborate canal networks. Arthur Cotton, a British general and irrigation engineer responsible for many of the nineteenth-century large-scale dam and canal projects in India, advanced the types of large-scale water storage in Asia that were already becoming the norm in the United States and Europe.

But even after gaining Independence, going big or going bust has been India's policy on water management. India ranks high when it comes to dam-building nations, third after the United States and China. In 1947, there were just 300 large dams in India. By 2014, 4,300 dams had been built[33] and another 500 were under construction.[34] In the Narmada Valley, a huge basin that includes the aquatic lifeline of Madhya Pradesh and its forty-one tributaries, development plans include thirty large dams, 135 medium dams and 3,000 small ones.[35] Resistance to these projects has been fierce.

For most of the twentieth century, the rapid expansion of surface and groundwater for irrigation, flood control and hydropower was seen as an unquestionably positive sign of human progress, but increasingly the limitations of this approach are being recognized. Underground aquifers that were thought to be bottomless, like the ones in Punjab and many other places around the world, are proving otherwise.

And there has been massive displacement of communities and fracturing of ecosystems in order to build mega dams or create major river diversions.

India's federal government seems unable or unwilling to use their authority to ensure that the country's environmental protections are enacted as large dam projects move forward.[36] It has pursued large-scale projects of unmanageable proportions, the Planning Commission tasked with overseeing interstate river projects neglects to do its job,[37] and projects that happen are wildly over-budget, often five times costlier than the original estimate.[38]

This aligns with the findings of a major study published in 2014 by Atif Ansar and others in the journal *Energy Policy*.[39] Looking at hundreds of large dams built around the world between 1934 and 2007, the great era of dam-building, they found that large dams routinely cost double what they are projected to cost, take nearly twice as long to build as scheduled, and have acted as major components of national debt crises in Turkey, Brazil, Mexico and the former Yugoslavia. The authors recommend 'agile energy alternatives' such as wind, solar and mini-hydropower instead.

The United States is slowly moving away from mega dams, and even dismantling older dams, the return of a wild Elwha River in Washington state the latest and largest of dozens of dismantlings occurring each year.[40]

Dams, writes Jacques Leslie in his book *Deep Water*, are at the heart of global conflicts over water scarcity, environmental degradation, biodiversity loss, development and globalization. And dams are a social justice issue, he argues, as they inordinately affect the lower caste and tribal peoples who frequently inhabit the less developed areas where dams are built.[41] He estimates that between twenty-one million and fifty-five million Indians have already been displaced by dams.[42]

Ramachandra Guha and Madhav Gadgil, in their book *The*

Use and Abuse of Nature, argue that large-scale development projects benefit India's 'omnivores,' the oligarchy and the upper echelon that most profit from an 'iron triangle' of state incentives, subsidies and technological interventions. They figure this is only about 16 per cent of the population. Those left out of the long-term benefits are the Indians they call 'ecosystem people', those whose land is fragmented or depleted, and the 'ecological refugees',[43] those who have been marginalized from their lands completely, often forced into day labour.

In addition to the socio-economic implications of large dams are the geophysical realities. The Himalayan mountain range region, with the most abundant water and most dam-building potential, lies in an area of tremendous seismic activity that makes large dams potentially dangerous. Like the grave consequences of other large-scale energy projects—consider the explosion of the Chernobyl nuclear power plant in the Ukraine, the meltdown of Fukushima Daiichi, Japan, after tsunami damage, and the Deepwater Horizon oil spill in the Gulf of Mexico that lasted for months—mega dams can lead to mega disasters should they fail. In India, the Morbi Dam failure of 1979, which was caused by heavy rains and flooding, killed an estimated fifteen thousand people.[44]

The Himalayas also span multiple nations, raising contentious concerns regarding international water treaties with neighbours such as Bangladesh, Bhutan and Nepal, not to mention Pakistan, with which India has been in conflict over water since Partition in 1947.[45] Already within India's borders, states squabble viciously over water rights.

Engineer-turned-journalist Hartosh Singh Bal, who experienced large dams as a child and felt it was 'perhaps the closest I will get to the Semitic God, awesome and terrifying', came to view large dams as a 'mistake' after years of reporting about them. 'Building a dam,' he wrote in his book *Waters Close*

Over Us: A Journey Along the Narmada, 'is a task that can be formulated and executed in a much more precise fashion than assessing its impact and trying to plan for it.'[46]

But dams, even the monstrous ones, seem almost quaint when considering the grandest water scheme of all: the Interlinking of Rivers (ILR) Project. The concept, first seriously considered in 2002, almost makes sense on paper. There is water in the 'surplus basins' of India's north-east, where the Himalayan meltwaters keep the Ganga, Brahmaputra and Meghna Rivers flowing. So why not just redirect the waters to the thirsty 'deficit basins' of the peninsula: the Cauvery, Krishna and Godavari Rivers in the south, where rainfall is comparably low.[47] Thirty concrete canals criss-crossing the country would create hyphenated offspring—Gandak-Ganga, Yamuna-Rajasthan, Farakka-Sundarbans, Koshi-Mechi, Ken-Betwa—and a deluge of accompanying dams and reservoirs. If enacted, it would be the largest water-transfer engineering feat on earth.

In 2002, the massive civil engineering undertaking had a price tag of Rs 7,60,000 crore[48] and plans for completion within an impossible sixteen years.[49] By late 2014, with little progress made on the project, the cost had risen (following the day-late and dollar-short script of mega dams), and it was estimated that more than half a million people would be displaced by the dozens of major proposed projects, which would also inundate formerly protected tiger reserves. Yet in late 2014, Prime Minister Modi, elected on a pro-development platform earlier in the year, urged the ILR Project forward, pushing for a fast track of mega irrigation projects and interlinking river schemes that could be enacted immediately.[50] By 2015, a task force was appointed to facilitate agreement within and between states to further speed up enactment. Modi had also quickly asked his administration to review cumbersome environmental regulations to see how they might be slimmed down.[51]

Whether the ILR or mega dams, with size comes bureaucracy.

The larger a water system, the greater the institution needed to fund, construct and manage it. Historically, public works departments replaced people, and not only did neighbours no longer come together to maintain the waters they depended upon, they were often forbidden from doing so. Instead, they paid taxes. Radically new ideas—that rivers could be owned, that land could be taken through eminent domain—evolved, and the usufruct that embodies the spirit of the commons, in which all villagers had rights to the use of trees, plants and animals of their shared lands, withered.[52]

The official creation of 'public lands' basically had the reverse effect. In 1945, the government set up the Central Waterways, Irrigation and Navigation Commission (CWINC) to survey and investigate water resources, partly in response to acute food shortages and the Bengal famine during World War II,[53] and by the time of Independence, state control of water was complete right down to inclusion in the new constitution.

'People started to believe that everything would be done by the government,' Kanhaiya had told me. 'Because the land became communal land, if they did anything on it—like make a johad—it could be said to be illegal and they could get in trouble.' The people of Bhaonta had been hassled by bureaucrats in the years leading up to international awards for their technically illegal efforts at dam-building in the common spaces, though it was the same work they would have been obligated, as citizens of their local communities, to do two hundred years earlier, with the full support of their ruling royalty.

The troubled state of India's water today—with dead and dying rivers, depleted aquifers and lakes suffocating from pollution—is a legacy of a government that has tried, yet ultimately failed, to get adequate water to her citizens. Over the past 150 years, Indians have been shut out from the process of caring for the water they depend upon. But in Rajasthan and elsewhere, people are reclaiming control.

THE RAINMAN OF RAJASTHAN

The Rainman of Rajasthan, the man who inspired Kanhaiya and had been the initial person to bring water back to the Alwar district and beyond, had not contemplated water at all when he arrived in Bheekampura thirty years earlier. He also did not consider telling his wife that he'd moved from their home in Jaipur to the rural outpost while she was off at her natal home, her body full with their first child. Or that he'd quit his stable government job.[54]

Rajendra Singh was the son of a landowner, but he wanted no land. As a child he had been unduly influenced, it seemed, when a young Gandhian who had come to stay with his family in service to the community became like a brother to him. He finally had a chance to act out his dreams of pursuing a life that would make that man, and the Mahatma, proud. It took half a year after Rajendra's child was born for him to confess to his wife the new direction their lives were taking.

'My wife was very much angry!' he said. 'With a job with the Government of India, the security comes without work!' He clapped and extended his fingers like a starfish, the sound echoing off the light blue walls of the TBS office, which was spacious but sparse. 'You just enjoy and money comes! Money and money!'

Rajendra didn't want money. He wanted to help shape his country, still so new and formative, still so in need in 1984.

Eventually, Rajendra's wife came around and supported his work as the founder of Tarun Bharat Sangh, which means 'Young Indian Organization', though by the time I met Rajendra, his thick beard and hair had mostly gone grey. But his face remained unlined by time. He was relentless in his activism, but took time for daily massages and contemplative walks in the garden. His son Maulik began working at his father's side in 2009, after a stint working in New Delhi's financial sector, and in the following years, had a wife and child of his own. TBS, along with a network of activists from down the road in Rajasthan to the Himalayan hilltops, has worked on water issues, both local and national, for the last thirty years, from the Arvari River to the Ganga, as well as fought against mining and mega dams and struggled to save India's tigers from extinction. It was enough to earn Rajendra the prestigious Ramon Magsaysay Award, often considered Asia's Nobel Prize, in 2001. A few years after I met him, he would win the 2015 Stockholm Water Prize, the king of Sweden presenting the award to him. He was a man whom people paid attention to, so when he told me to put on my shoes so that we could go for a walk, I did.

I had worn nothing but sandals on my feet for months of travelling in India, and my toes felt claustrophobic within the confines of my hiking shoes. But closed shoes seemed a good idea as Rajendra led me across the peaceful grounds of TBS, through a gate fastened shut with a piece of loose wire, and into the pokey underbrush of the forest beyond the buildings. In his fifties and fit, Rajendra, like Kanhaiya, wanted to show me his first johad, located a little over a mile away in Gopalpura village.

'There was nothing here when we came,' Rajendra said, of the 50 acres of land TBS acquired in 1988, which had only the buildings of a former hospital. The compound had the open, sprawling feel of Rajasthani structures, homes within homes, with courtyards and corridors that evoked my recurring dreams of discovering hidden parts of houses, tucked around corners and stumbled upon through low doorways.

There were 'no trees, no plants, no water', he said, pausing and digging the toe of his shoe into the dry layer of rocks and dirt to show me how little soil there was. But we stood in a forest thick with a variety of trees and the wide shoots of bamboo. Tiny leaves from khejri trees drifted down from above, and cows wandered freely through the shade. A bull, white with a brown-tinted hump, was lashed to a tree on the other side of a makeshift fence made of thorny branches.

Standing in the young forest reminded me of a favourite short story, 'The Man Who Planted Trees', an allegory by French writer Jean Giono, in which a solitary shepherd spends his life planting acorns, and the previously waterless, denuded landscape slowly comes back to life.

Rajendra moved quickly as we passed through a second gate where he grabbed a bamboo staff—a walking stick as well as a precautionary instrument of defence. We were adjacent to Sariska Tiger Reserve, once a place for hunting that now was a protected wildlife reserve. There were leopards and tigers in the area, he mentioned as we headed out in the late afternoon. We followed the path of a water hose on the ground, and Rajendra bent to adjust it at one point, so a leak would send its spray on a newly planted sapling. Further on, at the end of the hose's reach, Kanhaiya and another TBS staff member were filling up buckets and hefting them another 50 feet to water more new trees of an ever-expanding forest. Rajendra took a bucket and joined in for a few laps, watering multi-stemmed jatrophas and thorny khejris, their willow-like leaves folding into themselves as the day's light dimmed.

We continued through the future forest, ducking through barbed branches as burrs collected on our clothing. Shakira, one of a pair of sleek black dogs that belonged to TBS, was accompanying us, and she bounded with unlimited young energy along the path in front of us. It felt like leopard country—rocky, scrubby and open. A half-dozen nilgai looked at us from a hill a few hundred yards away before they moved away fluidly. Shakira watched them, snout lifted in the air.

As we neared Rajendra's johad, the land rose to a point, and he paused dramatically. Dark hunks of granite protruded from the earth, and he sat down on one. It was there at that rock that the direction of Rajendra Singh's life transformed, he told me. When he had quit his job, he'd come to rural Rajasthan with three friends. Inspired by the legacy of Gandhi, they wanted to help by setting up a hospital or perhaps a school. Then he met an elderly villager, who told him he and his well-intentioned friends were wasting their time.

'This is where Mangu Meena was waiting for me,' he said. 'Right here, on this rock.'

'"Forget about your mission of education or health care," the man named Mangu Meena told me,' Rajendra said. According to Mangu Meena, what the people of Alwar needed was water. It was all for nothing if they had no water. 'Dig tanks and build johads,' Mangu Meena had told the young Rajendra. 'You will get results.'

As Rajendra told me his story, a man appeared from the empty landscape, cresting the hill where Rajendra sat and I stood. He wore a bright striped shirt, and I learned that he was Mangu's nephew, who was just a small boy when Rajendra had met Mangu that day. They greeted each other as the nephew continued on his way, and Rajendra resumed his story.

'There are two hills where all the rain comes,' he said, as he stood and we began walking again, 'and Mangu-ji knew just where to put a johad to catch it. He showed me the vertical fracture in the hill and said, 'Put it there.' Before 1990, they had to walk to another village, 6 kilometres away, to get water.'

Only Rajendra rose to Mangu Meena's challenge. His friends left, their stint of youthful activism over. Rajendra spent the next three years living on rice and dal. With nothing but earth and stone and sweat, he built the johad we were now heading towards.

We moved in silence for a few moments.

'Do you think it takes an outside force to initiate change?' I asked him. In all my travels in India—in all my travels everywhere—I had been trying to understand what triggers change. When does it originate from within a community, and when does something—a person, an idea, outside news—need to infiltrate from elsewhere? What constitutes the seed of an idea and what are the necessary forces needed to make it stir out of dormancy, to sprout with life?

Rajendra thought about his answer for a moment before saying he did think change came from the outside. But thinking about his story of his meeting on the rocky trail with Mangu Meena, the elder who was very much of a place, and the young Rajendra, an outsider, it was clear it takes both.

Three young girls stood on the edge of Rajendra's johad as we approached, and they broke into grins, vying to get in the shots I was taking with my camera. Rajendra bent down at the water's edge and scooped up a handful of water. Algae hung suspended in the liquid, and underwater creatures surfaced just long enough to leave rippling rings in their wake. The basin was like a large pond, broad and shallow, and I tried to imagine a youthful Rajendra, shovel in hand, as I stood on its bank. He began the work, alone, in November 1985, and worked every day, he told me, eight to ten hours per day, for three years to finish it. When it was done, the johad was 260 feet long and 56 feet thick at the base. Mangu Meena cheered him on while most of the rest of the locals thought he was crazy. Each time the monsoon came during those first years of building, the johad would hold more and more water. In the third year, the wells in the vicinity of the johad replenished. His critics, he said, were converted.

Leaving the johad, we stopped by a nearby open well that was sculpted around a spring, enclosed partly with stone walls and partly with a natural earth embankment. He pointed out two pumps that reached into its depths, proving that it was a strong enough well to handle multiple siphonings. The water level of the well didn't rise higher than fifteen feet below the ground we stood on, but during the rains, it flowed right over the well's stone walls and into the fields behind us.

As the sun set and my mind returned to leopards, we turned to head back to TBS, Shakira still running circles around us. Just before we came to the compound wall, we crossed the agricultural fields that are part of TBS's land, where they grow the bulk of the food they consume. A man worked in a field, channelling water pumped from a nearby well to the bare ground in preparation for planting. There was a break in one of the irrigation canals, with water pouring out to the side, and Rajendra brought the man's attention to the breach. Another field, already planted, had rows of tiny bright green mustard plants poking up in neat lines.

As we crossed back over stone walls and through the gates, I asked what made him stay when his friends decided to leave Alwar. He wouldn't answer. I asked a few times, and he walked on in happy silence, ignoring me.

'What made you stay?' I asked again. Still he wouldn't answer.

'Did you know that Mangu Meena was right?' I asked, leading the witness.

'Yes!' he responded with sudden enthusiasm. 'I knew that he was right. Without water you have nothing! Without water there is no life.'

We returned through the grove of freshly watered saplings, a forest waiting to happen, leaving the imprints of our shoes in the mud, and arrived back at the compound just in time for dinner.

EPHEMERAL RIVERS

A week after Kanhaiya Lal Gurjar first took me to his village of Bhaonta, we returned to spend the night. Kanhaiya wanted me to meet a Bhaonta elder named Dhanna Gurjar.[55] Darkness was complete but for a three-quarter waxing moon as Kanhaiya led me to into a covered outdoor space in front of a house, where an aged man sat on a charpoy. One small oil lamp burned, casting a copper glow on his hollowed cheeks. He sat on his haunches, thighs pressed flush against his torso, kneecaps to shoulders, swathed in a wool blanket that left only his head, capped with a gauzy swath of fabric, exposed. He seemed a head bereft of body in the dim illumination, barely corporal. He greeted me as I sat on a charpoy across from him.

Dhanna's son Rameshwar came and sat behind me with a flashlight and shone it on us, so that everything was suddenly cast into a chiaroscuro, either blindingly lit or completely obscured. I couldn't see the man bearing the brilliant light, but made out the figures of children and others who had gathered around us. With the brightness, I saw Dhanna's eyes were clouded over with cataracts. The darkness and light meant little to him. He was eighty years old and the former president of the local village council, the Gram Sabha. Through translators, he told me his story.

'Before the water conservation efforts, I used to work in Delhi at the Lahori Gate, loading and unloading lorries,' he said. 'I went when I was thirteen years old, and was there for around fifteen years, starting before Independence, when Gandhi was still alive.' He told the story of his life through a time frame marked by events, not calendar months or years. Dhanna had been in Delhi when the Mahatma was still alive. He was there when Indira married Feroze Gandhi.

In the 1960s, Dhanna returned to his village of Bhaonta, hoping to raise animals, but there was no water for them, and the plants needed for animal fodder wouldn't grow in the drought. It was 'very, very difficult', he said. Before the johads, Bhaonta had thirty wells, but only four or five contained water. They provided enough water for the most basic human consumption, but there was not enough for the fields, or for farm animals.

One hand emerged from the blanket, clutching a pipe, which he scraped out as he talked, repacking it with tobacco he retrieved from a tin on a rough shelf to his right, his hands making the motions from memory.

'When there was no water, there was no agriculture. No water means no agriculture, means nothing,' he said. Monsoon rains were not enough. The villagers would sometimes try to make johads like the ones of the past, but they never made them strong enough, and when the rains came, they washed away. Then Dhanna and others learned about Rajendra Singh and Tarun Bharat Sangh and urged the village committee to seek their help.

'The elders told TBS about the village problems, how everyone had to leave to the city to find work,' he said, lighting his pipe. Rajendra Singh told them that TBS could help them make dams, strong dams that would hold fast, but it wasn't a charity. Each village had to contribute their labour, only then would TBS come.

The consensus to work with TBS did not always come easily.

Villagers would have discussions resolving community matters that went on late into the night, Dhanna recalled. 'They'd go on until 2 or 3 a.m. We'd visit each and every house to make them aware and teach them and make them understand the need for water, and why we should have check dams and johads to bring the water back.' Sometimes, talk wasn't enough to persuade a recalcitrant neighbour.

'If even after that some person didn't agree, we'd sit and demonstrate at their home until they agreed. It took some convincing for some people. Sometimes two to three people would be interested but one person would be resistant, not wanting to give the labour or something. So we'd just keep on talking and keep on talking, to make them understand.'

Dhanna represented one link in a network of water management that was created under the guidance of TBS. In addition to building the physical infrastructure with johads, TBS also helped establish the Arvari Sansad, the River Arvari Parliament, to maintain the watershed. Though the Parliament has no binding authority, it is composed of two representatives each from seventy-two villages,[56] giving everyone in the river's basin a democratic voice. It also has the kind of say that comes from being willing to demonstrate at someone's home until they arrive at an alternative point of view, a form of communal self-discipline, coercion for the common good. The Parliament has created a list of rules that these social mores maintain, covering everything from what crops are best to grow (barley and bajra) to what is forbidden (water-intensive sugar cane and rice paddy). It restricts when water can be extracted from the river, which is off-limits from the Holi holiday in the spring until monsoon rains arrive in July. According to Parliament decree, there shall be no bore wells.[57]

Conflict still emerges. In just one instance that Rajendra Singh had told me about, TBS constructed a dam on the Ruparel River in Alwar, and Bharatpur, a nearby city, said it had taken their water, though Alwar insisted that they were only containing water that had fallen on their lands. The Government of Rajasthan intervened and ruled in favour of Bharatpur, and TBS had to contest the decision to the high court, where it won in 2006.

Dhanna the elder leaned to the far side of the charpoy and spat on to the ground, then continued to tell me what I heard from the others. It took some time, but after the johads were constructed, the monsoon rains would pause in Bhaonta—a healing hiatus for the land—on their journey to the Yamuna, the Ganga and the Bay of Bengal beyond. The rain seeped into the earth in the pools behind the confining structures of the johads and found its way to the water table. Dhanna told of a time when farm irrigation began again, and agricultural production increased three- or fourfold in the years that followed.

'The money came,' he said. 'We started sending our children to schools. Before that, our houses were very ... not pukka houses, just mud. But after the money came, we made our houses pukka, and proper food was available. Now, after water conservation, each and every well has plenty of water.'

The steady high pitch of a cricket sounded in contrast to Dhanna's gravelly voice, which was soft with age.

Behind me, his son Rameshwar spoke up. He pointed out that if the water situation was dire back when his father was young, imagine what it would be like now, how much worse, given that the population had doubled.

Kanhaiya pulled out a heavy book, and showed me the Gram Sabha records from a decade earlier, with notes and signatures and thumbprints as well as comments from other visitors—a representative from the United Nations World Food Programme from South Africa, from Water Stewards Network, from Karnataka Water Development. Like me, they had come, from all over the world, to learn about what had happened in Alwar district.

After more than half a century of dominating water control, the government is beginning to delegate some of its authority back to its citizenry in the form of Water User Associations (WUAs). Several states have formed these groups to encourage active local involvement in maintenance and repair of water systems and more efficient use of water, recreating what existed more organically in pre-colonial times and replicating some of the work of the River Arvari Parliament. In some areas, the WUAs have been highly successful, such as in Andhra Pradesh and Tamil Nadu.[58] Still, in many states, WUAs lack ultimate control of water, which creates an arrangement that is vulnerable to conflict as national-scale projects such as the Interlinking of Rivers Project expand. Other critics say the WUAs remain exclusionary, more accountable to the state than to the people they serve.[59] Throughout the early 2000s, several states have also passed laws to protect waters—Andhra Pradesh, West Bengal, Himachal Pradesh and Chhattisgarh among them.[60]

Aside from these broader policy shifts, examples are emerging all over India of rain-men and rain-women who are finding ways to bring the precious and finite resource of clean water to their rural and urban landscapes. When Popatrao Pawar, the sole postgraduate in his village of Hiware Bazaar in Maharashtra, was pressured to become the sarpanch, or head of his village, instead of leaving to the city (by his friends; his family had urged him to leave), he succumbed. It was 1982, and the village was desolate, still unrecovered from a drought that had struck a decade earlier. Pawar decided that if he was going to stay in the village, he was going to make it a place worth staying in. One big step was closing the twenty-two liquor shops, but another came from recognizing, like the people of Bhaonta, that their agrarian life was impossible without water. He prioritized water conservation and management and rallied the citizens to create a massive rainwater harvesting programme. They built fifty-two earthen bunds, two percolation tanks, thirty-two stone bunds and nine check dams. Though in a rain shadow receiving an average of only fifteen inches of rain annually, Hiware Bazaar wells no longer lack for water. Between 1995 and 2012, the number of open wells with water increased threefold, and water was much closer to the surface. With water came wealth. More crops meant more income and more fodder for more cows. Milk production increased from 150 litres per day to 4,000 litres per day. In 2007, the village received the National Water Award for conservation, but it is the rupee millionaires the village brags about: Hiware Bazaar boasts sixty of them, and the water wealth has helped bring almost every single village family above the poverty line.[61]

'It is water that helped us become rich,' one farmer said.[62]

In other places, people are focusing on reforestation as a means to capture water. A couple of schoolteachers in Kasargod district in Kerala, on the Karnataka border, laboriously strategized with their students about how to plant trees on the school's barren 16 acres. Goats, thieves and the scorched earth conspired against them until they finally coaxed a few trees and plants to take root. More trees grew in their shade until, twenty years later, a multilayered forest of jackfruit, cashew, mango and gooseberry exists. Rainwater catchment atop the school has merely added to the new moist microclimate that surrounds it.[63]

Manav Adhikar Seva Samiti (MASS) constructs panigharas, another form of small water structures that collect rainwater as it flows downhill, in the eastern state of Odisha. For MASS, trusting local knowledge and rekindling the memory of water systems past has been the key.

'The villagers have better knowledge about water conservation in their areas than most engineers and we have a lot to learn from them,' Ranjan Panda, the founder of MASS, told the *Hindustan Times*. 'But the present system never seeks their views and sadly government officers dismiss them as illiterates.' Like TBS, the organization is expanding its work from local rainwater catchment to restoration of large rivers, in particular the Mahanadi, Odisha's largest river.[64]

In Madurai, Tamil Nadu, the Development of Humane Action (DHAN) Foundation has helped villages and towns restore derelict water tanks known as eris, crescent-shaped water tanks built into the slope of the land, and ooranis, rectangular tanks that are built up from the land to hold rainwater but keep animals out. Using similar principles as

Tarun Bharat Sangh, DHAN requires villages to contribute a third of the project cost, ensuring local buy-in and hopefully, in the process, decreasing the common mindset that the government will provide.[65]

In the new state of Telangana, establishing itself as distinct from Andhra Pradesh, more than forty-five thousand minor irrigation tanks constructed by Kakatiya kings are being restored, so that they can once again provide water for irrigation and household use.[66]

The international Food and Agricultural Organization (FAO) is collaborating with the Andhra Pradesh government's Farmer Managed Groundwater Systems Project, which educates farmers about aquifer basics, so that they can do their own monitoring and better understand how their water usage can affect the aquifer as a whole. Wells might seem like discrete structures but the water they tap into is a communal resource.

At the higher elevations in northern India, Ladakh in Jammu and Kashmir used to be fed by three glaciers that have vanished over the last three decades of warming temperatures. In the late 1980s, an engineer from Skara named Chewang Norphel noticed that a small stream that ran through a shaded poplar grove in his yard had frozen solid, while the water in the open sunny areas flowed freely.[67] He realized that if the pace of moving water was slowed, especially in shady spots, one could essentially grow glaciers. By building retaining structures out of the region's abundant stone, not unlike what Kanhaiya's fellow villagers built with mostly earth, the area's scant rainwater was slowed enough to freeze in place.[68] An artificial glacier could form in an age when the natural ones were on retreat.

The cities struggle, but some efforts have made headway there too. In Chennai, the saline waters of the Bay of Bengal were inundating the city's freshwater reserves. The people who live there, including my relatives, were going through increasingly complex and expensive means of securing water that they could drink and use in their homes along Elliot's Beach in the neighborhood of Besant Nagar. But in 2002, Shantha Sheela Nair, then secretary of the Water Supply and Municipal Administration, and popularly called the 'water woman', helped enact a mandatory rainwater catchment law across the city. The aim was to direct all the rainwater that falls on the roofs of the city directly back into the ground so it could recharge aquifers instead of draining off into the sea across the increasing mass of urban macadam. Supporting her efforts was Sekhar Raghavan, founder of the Rain Centre in Chennai and also a resident of Besant Nagar. There have been criticisms about compliance, both the rate and the quality of installed systems, but the groundwater levels are rising and temple tanks that had become cricket fields for young boys are once again filled with water.[69]

Chennai's Rain Centre is also actively working on waste and sanitation issues, because it's not just water coming in to homes and human bodies that conservation efforts are addressing, but also the fluids that carry off the effluent of our human lives. Waste needs somewhere to go and the nearest body of water usually suffers. In Kolkata and elsewhere, water contamination is the primary cause of ill health, from diarrhea to cholera and typhoid.[70] But in east Kolkata, restoration of wetlands, both natural and constructed, has helped process some of the city's waste, as well as create an ecosystem that supports biodiversity and aquaculture. The work earned recognition under the Convention on Wetlands of International Importance, better known as Ramsar.

The accounts I was hearing from across the continent were like wells coming back to life, like the seeds of the holy fig fruit—sprouting, hinting that there were still viable ways to quench the thirst of this fast-growing nation.

Everything I saw and heard about in Alwar was encouraging, but what did the science say? In the tiny kitchen of TBS, as I helped slip the papery skin off small cloves of home-grown garlic with the staff, they kept mentioning someone named Claire. No one could remember her last name, but she was from Australia, and she'd stayed with them for a year, maybe longer, researching their work.

I eventually tracked down Claire Glendenning, and we met at Khan Market's Café Turtle in New Delhi to talk about what she'd found in Alwar district.[71] Her work as a hydrologist started when she studied water recharge on cotton farms in Queensland, and that eventually led her to Tarun Bharat Sangh in Rajasthan. She too had heard about the accomplishments there, and she stayed through the monsoon seasons of 2007 and 2008 researching the recovered water systems of the Arvari watershed for her doctoral dissertation.[72]

Part of Claire's objective was to explore the arguments of critics[73] of small-scale water harvesting who argue that systems like the one in Bhaonta, when lacking consideration of the local hydrology, topography and agroecology, could end up reducing inflows into downriver reservoirs, creating local benefits but watershed disruption.[74] It's the same argument behind prohibitions of rainwater harvesting in US states such as Colorado and Washington.[75]

Over cups of tea, Claire told me about her research. She found that the rainwater harvesting in the basin had a large impact on the groundwater supply, just as the villagers had been telling me. The data showed the johads were capturing 7 per cent recharge, a substantial amount, and that water captured in a dam would pop up downhill, sometimes even a mile away.[76]

Claire did find limitations: an area could build only so many structures before their overall efficacy began to decrease—there was only so much water to be caught, after all[77]—but harvesting rainwater did help areas ride out periods of drought.

She also found that while there was some decrease in downstream flows when johads were catching the rainwater upstream, there was also a decrease in land erosion, especially as villages took ownership of their watersheds and actively reforested around the dams that they built.

Claire and I finished our teas and talk of water, two women from two parts of the world so far from where we sat, each facing their own stories of water shortage. As we stood to go, I called the Arvari a perennial river, and she corrected me. 'TBS says the Arvari is a perennial river,' she said, 'but it is not. It is a semi-arid ephemeral river.' She was in the southern part of the Alwar basin and it never flowed above ground for her two years there, she said.

It was a hydrologist's distinction, but the word 'ephemeral' stuck, forbiddingly.

Rivers and seas were once the circulatory systems of nations and pre-nations. Few humans lived far from their banks and shores. Even today, the majority of humanity resides on a shore or the banks of a river. Yet these aquatic life forces have been transformed by our presence. Sea levels are rising in some places, exacerbated by climate change, while rivers vanish in others. Both phenomena push people out. In the United States, the Rio Grande no longer reaches its former delta at the Sea of Cortez, the water dammed and diverted, siphoned for cities and agriculture. In India too the rivers that remain are dying. Faith in the goddesses many of the rivers are named after—the goddess Ganga who was falling from heaven until Shiva caught her in his locks of hair, or the goddess Yamuna, Lord Krishna's lover and twin sister of Yama, lord of death[78]—are not powerful enough to process the filth that has been poured into them: industrial pollutants, human sewage, agricultural run-off and more. Waterless rivers have become scoured courses of gravel, bereft of life, reincarnated as rock and sand quarries.

Environmental crises may have been at the root of many a great civilization's fall: animals hunted out, forests depleted. But drought has been the most implacable foe of all. People move to find water, and to avoid it. Looking forward, biologist Dr Hefin Jones from Cardiff University estimates that India will have around thirty million displaced environmental migrants within fifty years, counting not just internal displacement but an increasing influx from Bangladesh, the immigrants increasing on par with sea level rise.[79]

Looking back, a team of geologists, geomorphologists, archaeologists and mathematicians set out to unearth why the Harappan civilization—the great Indus Valley ancient civilization that once existed across northern India and Pakistan around 2000 BCE, with indoor plumbing, gridded streets and a thriving intellectual life—lasted two thousand years yet suddenly vanished.[80] The researchers concluded that as the monsoons lost their strength and precipitation patterns shifted, the Harappan were forced to leave their lands. The civilization edged south to the Ganga River basin.

Maybe the Sarasvati River, named after the goddess of knowledge and described in the Nadistuti hymn of the Rig Veda, was once the river that ran through the land of the former Harappan. Maybe she was once as grand as claimed, a thousand miles long, and maybe she lost the monsoon rains that once fed her, transforming her into a myth and memory.[81]

Ephemeral. A poetic word. It should conjure morning mist and the redolence of flowers in bloom, but it can apply to greater things too, rivers and even entire civilizations.

POWER FROM THE PEOPLE

Back in Bhaonta, we were hungry and knew the fresh maize rotis back at Kanhaiya's house would satisfy. After we took leave of the blind elder Dhanna Gurjar, we returned to Kanhaiya's pink concrete house for the night. I joined his wife Rama Devi and some of the children in the kitchen as one daughter,[82] Lada, made roti. The end of a large log burned in the fire, and smoke filled the space, hovering politely above our heads as we sat on the floor. Lada patted out roti after roti, each one wider than the span of her hand, searing each one on the hot curved metal of a griddle and then propping it against the edge of the mud stove facing the flames for one last tasty singe. Then she'd slather it with ghee before serving it with a side of cauliflower. There was nothing extraneous in the kitchen: a few tins of food and an army canteen. Kanhaiya's family purchased little, Rama said, just channa and sugar and salt, subsisting on what they grew for the most part. Their piece of land gave them wheat and some vegetables—carrot, tomato, cauliflower, bottle gourd—and the maize that made the delicious rotis we ate.

Rama Devi was enjoying the cooking break her eldest daughter provided, and another daughter helped. Two sons sat on a charpoy in the corner and watched. Rama Devi told me about the village she came from, not far away in Sariska Tiger Reserve, long before the park's boundaries were drawn. Her village, Kraska, was perched on the top of a hill, behind a Hanuman temple, she said, and water had been scarce. They had a drinking water shortage for all but two to three months of the year and would have to travel a mile or so downhill to get water, carrying it back up.

'It was a big amount of time and labour,' she told me, her children translating.

TBS had helped build johads there and now her family had become 'water rich', she said. 'Now we have five big johads right on the top of the hill. They're so big! Now all year there is water there. Earlier, only boys went to school, and girls would work with their mothers to get water. But now girls are going to school. This is because of the water only.'

And the women, I asked, what were they doing with the extra time?

'Their quality of life has improved without the labour of fetching water,' she said. 'There is more time to care for children and take care of the family.'

Lada silently patted out the next roti and placed it near the open flames.

When I met Claire Glendenning, I asked her what she thought of TBS's claims about how much their work benefitted women, having spent so much time in the area.

'I didn't see a big difference in the level of drudgery in women's lives,' she had said, 'even with water. The patriarchal society is still so strong that the girls might stay in school but they only go to local schools, while the boys are sent to boarding schools. It is still the women's role to prepare all the food, collect the firewood, and cook over an open fire.'

Water, in India, and many other places around the world, is a women's issue. Women and girls circle around wells and water pumps with brass or plastic vessels balanced upon their heads. Water is where their day begins, making chai, and ends, washing dinner dishes. Claire's words echoed what I'd seen in my travels. Women, even in the water-abundant Bhaonta, were up before dawn. The next morning I would awake in Kanhaiya's house, and stand on the front porch, watching his wife milk the water buffalo in the dark while his daughter laboriously worked a wooden butter churn in the corner of the kitchen. Next door, a man would sit on his porch, lighting up his hookah, the flame of the match illuminating his face in the predawn darkness.

After dinner, Kanhaiya and I sat in the front room of the house, half open to the cool night air, more dormitory than living room with half a dozen charpoys. The tall walls were painted a pale blue—only the sooty kitchen walls remained unplastered—with newspaper cut-outs for decoration. Kanhaiya wanted to talk more about governance, about the role of the Gram Sabhas and the Parliament. He wanted me to understand how important it was that every single family in the village was involved.

TBS's initial goal had been to spur village buy-in and emotional investment, but something unexpected happened when they worked together on these communal projects, Kanhaiya said.

Each worked together in labour, he said. It had definitely made a positive impact decreasing the differences between the castes.

It was reminiscent of India's Midday Meal Scheme, the one that had played an unfortunate role in the poisoning of the Bihari schoolchildren, but that dependably feeds 120 million schoolchildren each day. These communal feedings ideally put kids of different castes next to each other as they eat, something that caste conventions normally restrict.

It was this social intermingling, along with communal decision making, that made the work of TBS in the village of Bhaonta seem like it was fostering much more than just the physical natural resource of water. It was cultivating something less tangible, but no less important.

'In this case it's common people in the lower strata who contribute to the decision making, which is not normal, hai na?' Kanhaiya said as we sat on the cots with only a dim electric light to illuminate us. 'Usually, the higher-up and the wealthy, they make the decisions. Now the common man is part of making the decision. This has made a big difference. It has given respect to lower-caste people because it gives them the power to participate and they don't usually have that.'

Kanhaiya secured his woolen shawl tightly around his shoulders as the night grew cooler. The crickets continued their nocturnal refrain.

He told me that the common people work on the sites together, making them more self-confident, with greater self-respect. 'And then that decreases caste tensions, hai na?' He insisted that there was no pushback from the upper castes.

Kanhaiya's voice quickened.

'Now the people spread the news about their work wherever they go. They're really proud of what they've done, hai na?' He told me with his own satisfaction about meeting Prince Charles when he came to Bhaonta in 2003 to see the water restoration, which is land restoration, which is community restoration. The prince had arrived in a helicopter that landed at the temple about a hundred metres from where we sat. He was accompanied by Camilla Parker Bowles, and it was a hot day—that year was unusually hot, Kanhaiya remembered—so Camilla didn't walk up to the dam, but Prince Charles did. Kanhaiya, a low-caste villager, had stood with Prince Charles at the very spot he'd taken me to see, at the large dam at the top of the hill. The prince had stood in their little village in a white suit, as the villagers choked him with garlands of marigolds and placed a turban on his head.[83]

A few years before that, on 28 March 2000, President K.R. Narayanan came to visit Bhaonta to present the Down to Earth Joseph C. John Award. Thirty thousand people, by one count, poured out to greet the president. It gave a particular pleasure to villagers who had been harassed by government officials in the early days of the dam-building, since impounding water flows in any surface stream is technically a state matter, removed from the realm of village power. The pride that lingered in Kanhaiya, and over all of Bhaonta, was palpable. In 2003 and 2004, the Arvari River itself was awarded the International Riverprize.[84]

'You could say, when someone asked where you were from, that you were from the village of Bhaonta, where Prince Charles came,' said Kanhaiya. 'And all this work has been done by illiterate people. It's like it sends a message to the world, that even if we might be illiterate, we are very, very knowledgeable in managing our natural resources.'

The Alwar district of Rajasthan demonstrated how much water and power could be captured by a shovelful of dirt. It wasn't the only answer for India's water woes, but it seemed to be part of an answer, one that depended as much on geography and geology and sociology as on technology and economics.[85] A billion-plus thirsty people required more than a single answer, such as a continent-wide river-linking scheme. They needed a million little answers, rooftop catchments and dams they could construct in their own backyards and the planting of tree after tree that could possess the waters that came.

The debate will continue to play out, but Kanhaiya's words sounded prophetic and powerful.

'Before, an engineer might come to see about a dam, and the villagers would treat him like a god or something,' Kanhaiya said. 'But now they have done this work themselves. They alone have figured out how to build these dams, and they realize that it's not rocket science. And that makes them think about what other things they might be able to figure out on their own and question what things they wait for government to do for them.

'They say, We can do this. Hai na?'

3

AGNI

FIRE

Failure is an opportunity.[1]

—Lao-tzu

ETERNAL FLAME

In a village in Maharashtra, on Christmas morning, Seema Dattabay Kolekar squatted down in her kitchen and struck a match.[2] Primatologist Richard Wrangham once declared that humans are 'creatures of the flame'. Some have argued that it was this signature discovery—how to maintain and control fire—that caused our brains to grow and turned our species into what it is today. In Sanskrit, Agni (related to ignis in Latin and the English word ignite) is the name of the Hindu god of fire. He can take the form of the sun, of lightning and of the flame at the family hearth.

Seema set the match to some wood she'd brought in from the fields and it sent a last wisp of smoke up and extinguished itself as the wood in her mud stove caught. Smoke filled the room. Once the flames established themselves, the smoke slackened but didn't cease. A curled grey cloud climbed up walls already blackened by soot and seeped languidly through the open door and the one small window up high in the room. The morning light that slanted in was made solid as it struck the swirl of smoke. Seema stood up to reach for a tin of sorghum flour on a shelf, her brow pinched in the middle as though perpetually squinting into a very bright light, though the room was dark and cavelike. The air was clearer at low altitudes, near the padded dirt floor where we returned to sit.

Seema's chulha, a pan-Indian term for a variety of traditional biomass cookstoves, was shaped like a double horseshoe. Two cooking spaces sat side by side on the floor, crafted from mud like the plaster that coated the rough brick walls of the kitchen. The area was small—two steps across in one direction, three in the other. Seema fed pieces of split wood into the stove as she began to prepare breakfast for her waiting family.

This morning ritual is not confined to Seema's village of Nandal near the small city of Phaltan, in the Satara district of the state of Maharashtra. Across India, homemakers shake off sleep and set to arranging sticks, dried cow-dung patties, or waste from farm fields into a pyre. There is water to be heated; there are families to be fed.

Worldwide, two billion seven hundred million human bodies depend on cookstoves for food and warmth every day. Somewhere between a third and a half of the world's population is burning biomass, carbon-based materials, such as wood, dung, field waste and charcoal, that derive from living matter. The wealthier half of the world depends mostly on fossil fuels such as oil, gasoline, natural gas and liquefied petroleum gas (LPG). These powerhouses of energy exact environmental costs in extraction, but at the stove they burn much more cleanly than biomass.

In India, a nation that is rapidly developing in so many ways, 160 million households—more than two-thirds of the population—still rely on carbon-based fuel for their primary energy source.[3] Even as they pack into cities, many are not leaving the village fires behind. A quarter of those living in urban centers still use biomass. In villages, nearly nine out of ten homes are dependent on them in some way.[4]

'This is not going away,' said Dr Kirk R. Smith, an environmental health scientist at University of California, Berkeley, who has studied the health impact of cookstoves in India and elsewhere for decades.[5] Globally, the percentage of people who use biomass has slowly and steadily decreased over the last three decades, but because the population has been rising so quickly, he said, more people than ever are using these solid fuels.[6]

The impact on the climate and human health caused by the smoky emissions from household fires is devastating. A global health study released in 2014 estimated that household air pollution from cookstoves causes over four million premature deaths annually,[7] with more than a quarter of those in India.[8] The smoke from these fires contains dark particles that absorb sunlight and hasten glacial melting, and the soot has contributed to the formation of a brown cloud over South Asia so expansive that it is visible from outer space.[9]

Environmental organizations, development groups and governments have been grappling with cookstove pollution for decades, but momentum is picking up thanks to the Global Alliance for Clean Cookstoves, a far-reaching public–private partnership launched in 2010 by former US Secretary of State Hillary Clinton and other leading international figures. The Alliance aims to convince one hundred million households to adopt clean cookstoves by 2020.[10]

But what is a clean biomass cookstove? Traditional biomass stoves are little more than three stones to set a pot upon, with an open fire below. One could use bricks or broken bits of concrete instead of stones, but mud and clay stoves like the one that Seema used, are a little more permanent. For a hundred rupees, a potter will sell you a simple combustion chamber that you could encase in mud, and it might last a couple of years. All that is needed is a place to build a fire below and a place to rest a pot above.

An improved or clean cookstove is designed to use less fuel and burn cleaner. During the 1970s and 1980s, these devices were as simple as a chulha with a chimney. But there is now a whole new generation of clean cookstove engineers working from Colorado to Chennai, Bengalaru to Beijing, who are attempting to break new ground. There are high-tech fan-driven gasifier stoves by Philips. The Oorja burns pelletized field waste and uses a rechargeable battery pack to run a fan.[11] Brooklyn-based BioLite uses a thermoelectric coupler to power the fan (as well as a USB port that might make cellphone-wielding husbands more interested in their wives' cooking devices). Sleek units such as Envirofit and Prakti stoves have factory-made metal bodies and use natural draft and metered amounts of small-sized fuel to aim for a smoke-free fire in a design known as a Rocket Stove. Companies are using carbon credits (a financial instrument that offsets carbon produced in one place for a reduction elsewhere), corporate social responsibility initiatives, and microfinance loans to help fund these new endeavours. The question is: Are any of these attempts at clean cooking working?

Seema set a flat tawa griddle over the flames of her chulha. Her home had no gas fuel, though the heavy red cylinders of LPG were slowly starting to appear in her village of two thousand people, as they are in some other rural Indian areas, especially in the more developed south. Once, a couple of years earlier, Seema had tried an improved cookstove called the Bharat Laxmi. It was more elaborately designed than the clay stoves made by local potters. It had a combustion chamber made of heat-resistant concrete and an air-intake hole for improved draft, which were supposed to help the fire burn hotter and cleaner and reduce emissions. The individual pieces of the combustion chamber had been held together with wire and then enclosed by an outer layer of clay that made it look from the outside much like a traditional chulha.

Cummins, a global corporation that had been acquiring nearby village lands at a rapid clip to set up a megasite of industrial factories, had donated five hundred stoves for the villagers as part of a corporate social responsibility (CSR) campaign, and Appropriate Rural Technology Institute (ARTI), a well-regarded non-profit organization, had installed the stoves, which cost Rs 700, in village homes.

'We had a Laxmi,' Seema said, 'but it broke after one year.'

Seema adjusted her orange sari and opened the flour tin, sifting sorghum flour into a shallow copper vessel in front of her. She steadied the pan with a big toe on either side, her knees jutting up, silver toe rings encircling her second toes (a sign of her married status) and silver anklets shining in contrast to her dark skin. She added some water to the flour and began to knead the dough. Then she broke off a nub and patted it with a rapid repetitive motion into a perfectly thin, perfectly round bhakri, the traditional Maharashtran style of roti bread. She'd been adding pieces of wood to the fire between pats of the dough, and the tawa was perfectly hot when she slapped the bread on. Her hands winged between her tasks. The rhythm of her kneading was like a steady heartbeat, punctuated each time as her green bangles struck one another.

'Even when it worked,' she continued, telling me about the supposedly improved stove, 'we could only use it for vegetables. It wasn't hot enough for bhakri,' whose surface should be seared by the flame, creating tasty blackened blisters on the surface of the bread. The improved stove used less wood, she said, and she used it to cook vegetables, but she'd have the chulha burning simultaneously to make the bread the way they liked it. So Seema never noticed much of a difference in the smoke levels in her kitchen. When the Bharat Laxmi broke, she returned to the double chulha.

Seema and her sister-in-law Vanita were the cooks of the household and wives of the two Kolekar brothers. Seema's husband received the first bhakri, pulling off sections to eat with an eggplant curry and bit of thecha pickle, a freshly ground chilli and garlic paste. He was silent and subdued in the women's space of the kitchen. Seema's young son received the second bhakri. Then the grandfather entered, dressed completely in white but for a light checkered cloth draped over his Gandhi cap, warding off the morning chill that still hung in the air. Seema fixed the men tea, and the grandfather sat on his heels and sipped his from a small china cup. Seema returned to her cooking, with the sound of her bangles and the swish-swish of the bhakri-making as background.

I asked Seema and Vanita if they'd prefer to have LPG, and they answered yes immediately, but the grandfather, teacup in hand, balked.

'To spend such money!' he said. The husband, who had been eating in sullen silence, added that the bhakri tasted better on the chulha.

'I prefer the bhakri made on this,' Seema agreed. She shifted a stick in the fire, flipped a bhakri on the tawa with her fingers, and returned her hands to the pan between her toes to spin out the next lump of dough.

The men used the last of their bhakri to mop up the last of their curry from their stainless steel plates and left. Shifts of more children entered. Now that the men were gone, I asked Seema again about gas, or if she would spend the money for another improved stove.

'Sure, I would like another stove, but my husband would have to agree to spend that money,' she said. 'And, really, if we can get a potter to come make a chulha for fifty rupees, why should we spend a thousand on this other stove?'

In two short sentences, Seema had summed up many of the roadblocks that policymakers, staff of non-profit organizations, stove designers, energy visionaries, UN delegates, impact investors, foundation heads, and staffers from a slew of organizations—UNDP (United Nations Development Programme), EPA (Environmental Protection Agency), USAID

(United States Aid for International Development), GIZ (Deutsche Gesellschaft für Internationale Zusammenarbeit GmbH)—have been trying to surmount for forty years: economic limitations, cultural habits, gender inequality.

If you want two billion seven hundred million people to give up their traditional cookstoves and use something cleaner for them and the planet, there are a few requirements you must meet. The stove must be something people want. It must use the fuel they have at their disposal. It must make people's lives easier, not harder, and the time they spend cooking shorter, not longer. And it needs to make traditional food taste like it is supposed to taste (bhakri toasted just so). The stove has to burn better than a chulha, hotter and cleaner, emitting less carbon dioxide, carbon monoxide and particulates. It has to last a good long time to justify the price, and the price has to be damn close to free because if the 'bottom of the pyramid' population had any spare money it would likely be spent on a mobile phone, a refrigerator, a blender or a motorbike. So the cookstove needs to be efficient, desirable and cheap. It also has to be 'aspirational', a word I heard over and over. Whatever you do, don't tell your consumer that you've made something called Friend of the Poor. Who wants to be associated with that label? The Tata Nano was marketed as 'cheap'. It flopped.[12]

High prices and gas unavailability still keep the vast majority of families from switching away from biomass, which can often be collected for free. So those concerned about the health impacts of traditional stoves have pinned their hopes on the ability to create a clean or improved biomass cookstove.

Yet even with technological advancements, no one has yet created a biomass stove that is truly harmless to human and planetary health *and* affordable and desirable to the families who need them most. The crux of this cookstove conundrum is partly technological, but it is also deeply cultural. Despite the multidisciplinary approach now infusing the world of clean cookstoves (involving the fields of epidemiology, climatology, gender empowerment and global finance), consumers remain reluctant to warm to these devices.

Why is improving something as basic as a wood cookstove such an impossible task? And if it is, should we consider alternatives?

It is time to move beyond Palaeolithic methods that cause so much pollution, both inside homes and outside, says epidemiologist Kalpana Balakrishnan, director of the WHO Collaborating Centre for Occupational and Environmental Health at Sri Ramachandra University in Chennai. 'Forget about household air pollution,' she told me. 'If you want clean air anywhere, you don't want to be burning biomass.'[13]

While there are areas where fuel sources are scarce—the salt flats of Gujarat and the most densely populated spans of the Indo-Gangetic Plain—Indians don't struggle to find biomass fuel the same way that sub-Saharan Africans do. I saw evidence of fuel collection scattered across the landscape. There were neat piles of slender branches stacked outside homes in Karnataka. Walls were plastered with discs of drying cow dung mixed with hay in Bihar. In Punjab, I saw limber-limbed children scrambling up already pollarded trees to hack away at the remaining branches with a machete. In Tamil Nadu, a lone woman dragged a 20-foot trunk along a sandy path, her sari blowing in the sea breeze. Everywhere, there were women with loads of wood—with an average weight of 70 pounds—balanced upon their heads. These are necessary burdens given that the average India family requires 40 pounds of fuel every single day to cook the rice and simmer the curries, to roast the rotis and heat the bath water.[14]

Biomass might be renewable but it weighs heavily on the backs of those who must collect it. And those people are typically women. Women crush their spines, wear down their cartilage, and risk uterine prolapse from carrying wood. They make themselves vulnerable in isolated spots where wood is easier to gather and the threat of sexual violence is real. But when there are family lands nearby, women told me, hours spent collecting wood with other women was precious—a time to gossip and catch up. I heard of a toilet project in one village that failed simply because the women valued their walks into the fields too much. What other private moments did they have to share?

Seema and Vanita didn't have to go far to find their fuel. While Seema's husband is a builder in town, Vanita's husband works the family's agricultural lands, so the women can gather nearby wood easily.

'We cut from the tree and bring it back bit by bit each day,' Vanita said. The sisters-in-law spend about an hour in the fields dotted with acacias and other small trees.

It was just as R.D. Deshmukh, the founder of ARTI, had told me when I'd met him earlier in Pune. 'When there is agriculture, then the cookstove is there.'

On another day, in the same kitchen, Vanita tended the same chulha.[15] It was bright outside, and my eyes took a few minutes to adjust to the smoke and dimness. I entered standing, but had to squat down to catch my breath and keep my eyes from burning. Vanita roasted peanuts above a fire on one side of the chulha while eggs boiled in a pot on the other. Her eighteen-month-old son sat less than a foot away, propped up in front of the flames. Snot extended from nose to chin. His eyes followed the flickers of fire and his mother's hands as she prepared small portions of eggplant, garlic and onion, and made a paste from chilli and coriander. The spices mingled with the smell of woodsmoke. The smoke didn't seem to bother either Vanita or her son.

The newest health data paint a stark picture of the human health impacts of cooking with biomass. Household air pollution from biomass stoves poses the greatest health risk in the world after high blood pressure, alcohol use and tobacco,[16] and is the second worst risk for women and girls,[17] except in India. In India, it is the first.[18] But everyone is affected. More people die from the incremental, ongoing inhalation of smoke from fires they ignite in their own homes than from malaria, tuberculosis and HIV/AIDS combined.[19]

There are 'non-idealities in combustion', thermodynamic and kinetic constraints, writes Gautam Yadama, author of *Fires, Fuels and the Fate of 3 Billion*. Ideally, every last bit of carbonaceous material—the slivers of straw, the fibres of wood—would be oxidized when burned, leaving nothing but carbon dioxide and perhaps a mist of water vapour. But rarely if ever is the fuel so clean, or the combustion so complete. And even if it were, carbon dioxide contributes to climate change. Earth's atmosphere exceeded 400 parts per million (ppm) of carbon dioxide in the spring of 2013.[20] Many climate scientists think we should be closer to 350. The pre-industrial level, which lasted for at least a million years, was 275 ppm of carbon dioxide. And then humans began to burn coal, oil, gas and other fossil fuels faster than ever before. But traditional cookstoves and puja fires that proliferate across South Asia play a significant role as well.

Depending on the fuel, flames can release, in addition to carbon dioxide, sweet-smelling, sleep-inducing, leukemia-causing benzene,[21] the known human carcinogen[22] (and food preservative) formaldehyde,[23] and chloromethane,[24] which animal studies have shown to cause slow growth and brain damage even at low doses, damaging sperm in males, and causing pregnant females to lose their fetuses.[25] Other off-gases include sulphur dioxide[26,27] and polycyclic aromatic hydrocarbons, which encompass a hundred different chemicals that can affect the skin, liver and immune system.[28] The nitrogen oxide[29] that commonly swirls around us in the ambient air also concentrates around a cookstove, causing fluid build-up in the lungs of those nearby.[30] And of course there is odourless carbon monoxide,[31] at one time used as a form of execution. (A man would be placed in a sealed room with burning coals and slowly suffocated.)

'Soot' is the simple word for this tangled cloud of organic carbonaceous particles but the technical term is particulate matter (PM),[32] and it affects more people than any other pollutant on earth.[33] Having a biomass cookstove burning in the house is not unlike having a miniature volcano or forest fire aflame in the living room. Though they span a wide range, PMs are classified as coarse or fine, based on their aerodynamic diameter. The coarse ones are PM 10: smaller than ten micrograms (Mg) per cubic metre of air. Fine particulates are PM 2.5: smaller than 2.5 per cubic metre of air. Both are small and light enough to easily become airborne as aerosol emissions. These aerosol emissions (not to be confused with the stuff of spray cans—that's something different) are the wandering minstrels of particulate matter, and epidemiologists are finding them increasingly vexing. Yet Seema, like your average high-ranking public policymaker, rarely associates those smudges on the wall with something that might cling to the walls of her organs and cause harm. Whether the soot reaches Vanita's sniffling son just inches from the fire or travels aloft for miles—soot from India has been found in the Maldive Islands and on the Tibetan Plateau[34]—aerosol particles smaller than one hundred micrometres in size are inhalable. If they are smaller than ten micrometres, their journey can continue through the upper respiratory tract, past the phalanx, past the larynx, deeper into the trachea, and down into the lungs.[35] Particulates smaller than PM 2.5 are even more dangerous, slipping deeper into the peripheral regions of the bronchioles, and the tiny tips of the lung's branches, blocking oxygen's admittance into our bodies.[36]

Once in the body, smoke from biomass combustion is responsible for increasing incidents of disease, especially in women and children, including acute lower respiratory infections, chronic obstructive pulmonary disease, cardiovascular disease, pneumonia, cataracts and lung cancer.[37] Soot causes half of all fatal pneumonia cases in children under five.[38] Close to half a million of these fatalities occur in India.[39] Already, women and girls bear the brunt of the physical hardships and danger of collecting firewood outside the home, and within it they are exposed daily to huge amounts of pollutants. One study showed that Indian women cooking in solid-fuel-using households are exposed to a mean twenty-four-hour PM 2.5 concentration that is thirty times greater than the WHO indoor air quality guidelines.[40] Another study, of households using traditional cookstoves in the Indian state of Odisha, found that a fifth of the children had carbon monoxide levels of ten ppm, the same as that of a heavy smoker.[41]

While the developed world struggles to clean up its skies, and takes a smoke-free kitchen for granted, the developing world is contending with worsening air quality, indoors and out, that is killing people at a staggering rate.

'You know when awareness about this started?' asked Dr Kirk Smith. We were sitting in a cheerful cafe in Hauz Khas, New Delhi, talking about his work over the last thirty-two years on cookstoves in India. In his mid-sixties, he was wearing a tan patterned kurta and tortoise-shell glasses that hung around his neck from a thin loop. His substantial research has guided the findings of the Global Burden of Disease Study, the policies of the WHO, and the recommendations of the Intergovernmental Panel on Climate Change (IPCC). Recently it had earned him the Fulbright-Nehru Distinguished Chair, which brought him back to India on what he figured might be his eightieth trip to South Asia.

In answer to his question, I took a stab and guessed that awareness about the dangers of traditional cookstove smoke might have begun in the early 1980s. It was then that India's government first showed interest in improved stoves, and when engineers and tinkerers around the world were beginning to think about how to make wood burn more efficiently in household stoves, mostly inspired by concerns about deforestation.

Kirk shook his head. '1959,' he said, his gaze hovering over our sandwiches as he took another bite of his chicken.

'1959,' he repeated, pursing his lips. 'The director of the National Heart Institute noticed that she was treating all these young women,' he explained, leaning back in his chair, 'young women who were rural, non-smokers. Some with severe *cor pulmonale*,' an enlargement of the right ventricle of the heart, the body's response to increased pressure in the arteries of the lungs, which in turn is a response to low blood oxygen. This doctor, ninety-six and still living in New Delhi, had figured out more than half a century earlier that dung smoke was causing heart problems in young rural women.

Nearly two decades later, in the late 1970s, the question of the human health impact of biomasss stoves was still off the development radar. Smith was doing a postdoctoral research at the East-West Center, a US-funded non-profit organization focused on environmental health and nuclear safety, without a thought to cookstoves. After agreeing to head the energy programme there, but young and itching for adventure, he negotiated six weeks of travel in South Asia, through the Philippines, Thailand, Sri Lanka, Bangladesh, Nepal and India.

'What I saw of village life was a smoky situation,' he said of the trip, a defining moment of his life. 'I was trained in environmental health, but I couldn't find anything published about this. We did some back-of-the-envelope figurings based on the size of the kitchens we saw and carbon output from stoves and air pollution. The numbers were terrifying.'

So in 1981 he cobbled together funding from various sources, bought some air samplers and returned to India to see how the envelope estimates compared to field data. He and his team of researchers showed up in four small Gujarati villages in late November of that year and set to work. They hooked up air samplers to seventy different women as they cooked, took samples from surrounding rooms and the outside ambient air, and found one token male—a migrant labourer who was cooking his food in a sheet metal hut—to test as well. They recorded fuel use and the speed of wind while the air samplers collected data about individual exposure to emissions. Researchers tried to stay in a room where the roof hole was plugged, to simulate conditions during monsoon season, but couldn't stand it for more than a few seconds because of the smoke. Their findings basically supported Smith's back-of-the-envelope estimate. Cooks were spending two-and-a-half hours each day around fires that were emitting particulates at rates that far exceeded global standards, breathing in more pollutants than you'd find at a traffic stand in the middle of a large city. A woman cooking with a biomass stove might as well be smoking twenty packs of cigarettes a day, the study reported.

Smith immediately lost all interest in nuclear safety and shifted his research towards the pursuit of a clean cookstove that people wanted to use. Decades later, he was still looking.

I had travelled across north India for months under a perpetual layer of yellow, acrid 'fog' that did not look anything like innocuous mist. I had smelled dung burning in cookstoves in poor Bihari households as women stretched a handful of dal and greens into a meal and in wood fires in the kitchens of wealthy Punjabi farm estates, where multiple-course vegetarian meals were prepared from the bounty of the land. What had led me to Seema's village was the claim by ARTI, the non-profit organization that had distributed the improved Bharat Laxmi stoves, that Nandal was smoke-free because of its stove programme.

Though ARTI is based in Pune, it has a training centre a few hours outside the city that has served as its rural outpost since 1996, providing a hub of improved cookstoves and alternative energy technologies as well as a nursery for sugar cane seedlings, of which they sell one million per year. Their stoves are a slow-moving commodity. According to Narendra Zende, who runs the centre, fifty sales were considered a good month. On a visit,[42] he led me outside to their demonstration area, which included a gazebo filled with a dozen improved stoves designed by ARTI and others, a briquette press that turned sugar cane waste into a compact fuel source, and a biogas digester.

The demonstrations looked so good. There was the simple Bharat Laxmi stove along with other stoves that had double burners, some with chimneys and some without. One, based on a lauded stove design by Philips, had a fan. In another area was a metal contraption of pipes and valves nearly 30 feet long: a biogas digester that took methane-releasing manure and transformed it into clean gas that was then piped into the centre's kitchen. Billows of smoke rose from an oil drum that women had jammed full of sugar cane bagasse—a residue slow to decompose and worthless as animal fodder—and lit. The remains were pressed through a simple mechanical auger that produced briquettes that were little black energy bombs. For each hectare of sugar cane, ten tonnes of waste are generated, and the biochar that ARTI was promoting recycled waste that would normally be set uselessly on fire.[43]

From the ARTI centre, I travelled a half-hour to the village of Nandal, over roads that alternated between rubbled lanes that thrashed our small vehicle and a streamlined divided highway with the occasional errant driver speeding upstream against traffic. We passed corpses of dogs, shops whose walls were emblazoned with smiling LPG tanks advertising HP Gas, and schools with murals of kids on flying pencils. We drove past houses with chulhas burning out in front and satellite dishes up top. A man's axe glinted in the sun as he chopped off rough sections of wood from a downed tree.

The land buzzed with the activity of the sugar cane harvest. Labourers slashed at stalks twice their height, loading bullock carts until not one more cane would fit. The lumbering caravans then trekked along the shoulders of the road. Children peeked out of blue tarp tents of migrant labour camps set up in the bare spots between fields. While I was in search of smokeless stoves, there, beyond the village bounds, billows of smoke rose from acres upon acres of burning sugar cane fields.

Nandal was down a road flanked by tall stone walls topped with barbed wire. Behind the walls, Cummins factories manufactured high-horsepower engines and low-horsepower generators, power for people with means.

Those were not the people of Nandal. In this small figure-eight of a village, as well as other villages nearby, I stepped into house after house, and though many, like Seema, had received a free Bharat Laxmi stove just a few years earlier, I couldn't find a single stove still operating in any condition resembling what its designers had intended.

First of all, no one liked that air-intake tunnel built into the bottom of the stove. Women plugged it out of the reasonable fear that snakes or scorpions would mistake the narrow passage for a nice den. Some liked the stove, and used it regularly, until the wire holding the bricks together burnt out in weeks, months, or at most a year, leaving nothing to rest the pot on. When the bricks collapsed, people patched it up with mud, transforming the new and improved cookstove back into an old-fashioned chulha. Some claimed that, even when working, the high-tech stove used more wood, not less. Others, like Seema, complained that it wasn't hot enough to cook certain foods. A taller chimney allowed flames to fully combust, making it cleaner, but it also meant the flame that hit the bottom of the tawa wasn't hot enough to cook bhakri the way they liked. Husbands were universally not interested in the stove one way or the other.

In spite of the website declaring Nandal a smoke-free village, the project that Cummins paid for and ARTI oversaw was a dismal failure. I don't doubt that ARTI does valuable work, but its stove project was not working, even by its own admission. Two years after the Bharat Laxmi stoves were distributed, and six months before I arrived, ARTI staff returned to check in with some of Nandal's villagers. Zende showed me the survey results. More than half had removed the stoves from their homes, complaining about the broken wires. The survey said that just under 20 per cent, a figure that seemed generous from what I saw, were still intact and in use.

One surveyor reported comments from villagers: 'Not at all happy about Cummins giving this cheap stove.' The Bharat Laxmi was the bottom-of-the-line ARTI stove, with a single burner (when most people wanted two) and no chimney. They knew they were getting a freebie. 'Why are they cheating us by giving us things that break so early?' a respondent complained. 'Why don't they give something more substantial like LPG or toilets or jobs for [our] children?'

I did see some Bharat Laxmi stoves that were in pristine condition, but only because they were completely unused. At the opposite end of Nandal from Seema and Vanita's home lived Lata Kisan Khare. Her husband and her neighbours' husbands were all musicians, and a colourful miniature car was parked at the centre of the courtyard where their cluster of homes faced. Lata, like Seema, was also unconcerned about the particulates she was inhaling.

'The smoke doesn't bother me,' Lata told me, pointing to the traditional chulha that was still warm from breakfast. 'It just goes up and away.'[44]

Lata's kitchen was a makeshift space added to the front of her house, with a sheet of corrugated metal nailed between posts as a half-wall. A multicoloured goat was tethered nearby. The mud stove was on the ground, with a kerosene stove and a box of matches next to it. When I asked her about her Bharat Laxmi stove, she led me inside the finished part of the home, where drums and a trumpet hung from the walls and a low ceiling, and cornstalks dried in a corner. There was a Laxmi, perfectly intact and perfectly unused, with a few pieces of crumpled paper on top of it. Not one smudge of soot blemished the wall behind it. I asked her why she wasn't using it.

'It's better to use the one outside,' Lata explained sensibly, 'so there's no smoke in the house.'

Ask most women if the smoke is a problem and you'll receive a dismissive shrug, born of fatalistic acceptance perhaps that, with fire, there's smoke. But, of course, the smoke doesn't just go 'up and away'. The delineation between indoor and outdoor is amorphous at best in many developing countries, where mud stoves might be in the corner of a one-room home, but are just as likely to sit, like Lata's, sheltered under an eave on the outside wall, or in a temporary thatch shelter set aside from the main house. Researchers for the most recent Global Burden of Disease Study, a collaborative assessment of what kills and sickens humans worldwide, revised the term indoor air pollution (IAP) to household air pollution (HAP). Globally, nearly a fifth of outdoor air pollution originates not from outdoor sources such as tailpipes or factory smokestacks but from within the home.[45] In India, where smog levels rival those in China,[46] more than a quarter of the fine particulate matter in the ambient air originates from household cookstoves.[47] Homes, often considered a refuge from air pollution, instead turn into a deadly source of it when biomass stoves are in use.

Kalpana Balakrishnan, the epidemiologist who told me there was no good way to burn biomass cleanly, has found that even people in households that have transitioned to LPG and other cleaner fuel sources still have elevated pulmonary risk if their neighbours continue to cook with traditional fuels.[48] It's a variation of 'herd immunity', in which a community can avoid contagious diseases if a high enough percentage among them is vaccinated, but become vulnerable to outbreaks if not. The personal is the communal and, overhead, the communal skies are darkening.

Over South Asia a massive brown cloud is contributing to climate change, and Dr Veerabhadran Ramanathan, professor of climate science at the Scripps Institute of Oceanography and UNESCO chair professor at TERI University in New Delhi, believes that the smoke from cookstoves is its leading cause.[49]

When he was a young engineer in India, Ramanathan suffered through a tedious job tracing chlorofluorocarbon (CFC) leaks in a refrigerator factory before escaping to Bengaluru for graduate school. There, he dreamed of making it to the United States so that he could live a good life and drive 'big, fast cars'.[50] He wasn't exactly on a trajectory to be fighting climate change.

Ramanathan made it to the United States in 1970 and dabbled with studying the atmosphere of Venus before he came across ozone hole research under way at the National Center for Atmospheric Research in Boulder, Colorado. He began to compare the impact of CFCs like the ones he'd traced at the refrigerator factory to the greenhouse effect of carbon dioxide on the earth's climate.

Not all molecules are created equal, Ramanathan and others were discovering. In 1975, he authored a paper in *Science* showing that adding one molecule of CFC in the atmosphere would have the same greenhouse effect as adding ten thousand molecules of carbon dioxide.[51] Researchers increasingly revealed an array of greenhouse gases that were creating a chemical soup overhead: methane and nitrous oxide, stratospheric and tropospheric ozone and carbon tetrachloride among them. How all these chemicals interact with one another is something scientists are still trying to understand.

Curious, Ramanathan wanted to get into the skies over his Indian homeland to learn more. In 1995, he helped launch the Indian Ocean Experiment (INDOEX), a research project that would grow to include six drone aircraft and scientists from around the world. INDOEX's investigations found a haze that reached to the heavens, 2 miles thick, covering the Asian subcontinent and spilling across the waters of the Arabian Sea and the Bay of Bengal. This Asian brown cloud covered an area equal to the continental United States and was filled with a frightening quantity of black carbon. It was the atmosphere's version of the Great Pacific garbage patch, where currents have amassed a concentration of marine debris, mostly plastics.

In the summary of *Atmospheric Brown Clouds,* produced by the UN Environment Programme with Dr Ramanathan as lead author, photographs of cherubic carefree Asian children smiling amid the bright green shoots of a young rice crop sit uncomfortably at odds with the report's apocalyptic analysis. Soot and other aerosols are not just harmful to human health. They fundamentally threaten water and food security in Asia. The hovering cloud reduces the amount of sunlight hitting the earth's surface and disrupts the water cycle, thus lowering crop yields of staples such as rice and wheat by up to 50 per cent.[52]

The cloud also impacts the cycle of evaporation and precipitation that in India equates with the monsoon. Though never altogether dependable, monsoon patterns seem to be growing more erratic. Rains are needed to replenish the Himalayan snowpack, which feeds the Indus and the Ganga, the Brahmaputra, the Mekong and the Yangtze Rivers. Together these rivers feed a large percentage of the world's population; the Ganga River basin alone is home to 407 million people. Rainfall over the Indo-Gangetic Plain has already decreased by a fifth since the 1980s.[53] When the rain does come, it comes faster and harder, causing torrential flooding, like in Uttarakhand in the summer of 2013 and Kashmir in 2014. Also, when bits of black carbon, from cookstoves and elsewhere, settle on snow and glaciers at high elevations, they absorb sunlight and emit heat, thus speeding up the process of glacial melting at a time when elevated global temperatures are already depleting the planet's ice chests. The Gangotri glacier, which feeds the river Ganga, is retreating 115 feet each year. Should the glacier ever disappear, the holy Ganga would become seasonal.[54] Explain that to the gods.

'If you want to slow melting,' Pam Pearson told me over the phone one day, 'the best way with the hugest co-benefits is cookstoves,' meaning decrease their emissions or get rid of them altogether.[55] Pearson is the director of the non-profit organization, International Cryosphere Climate Initiative (ICCI), which focuses on the shrinking snow- and ice-covered portions of the planet. She is also one of the lead authors of the 2013 ICCI–World Bank report 'On Thin Ice'. This report lists cleaner cooking solutions as both its top priority in mitigating climate change impacting the cryosphere and the only proposed strategy that would have a significant effect on every continent on earth, including Antarctica.[56]

Likewise, a study published in the *Journal of Geophysical Research-Atmospheres*[57] showed that residential burning of solid fuels, including biomass and coal, accounts for a staggering 60–80 per cent of black carbon emissions in Africa and Asia, leaving black carbon second only to carbon dioxide in terms of its heating effect in the atmosphere.[58] This study also suggested that climate scientists are not adequately accounting in their modelling for the impact of black carbon, which could be almost three times worse than currently believed.[59] This research came to the same conclusion that Pearson did: getting rid of smoky cookstoves is one of the best ways to improve human health and address climate change.[60]

Ramanathan agrees, describing improving or replacing inefficient biomass stoves as a 'low-risk, cost-effective and high-reward option' for fighting climate change.[61] And the payback is immediate. Carbon dioxide lingers in the atmosphere, so even if we cut back on emissions the effects of carbon dioxide from years past will continue to be felt for decades or even centuries to come. But Ramanathan's data shows that, once black carbon is removed, an improved atmospheric effect is noticeable within months.[62]

One complicating factor is that the sooty ash from cookstoves definitely contains heat-absorbing black carbon but also includes light-coloured aerosols that reflect solar radiation, helping to cool the air as well as Earth's surface. When Mount Pinatubo erupted in the Philippines in 1991, it sent up fair sulphate particles that cooled the atmosphere for the next two years.[63] One terrifying notion is that these aerosols could actually be masking some of the warming effects of carbon dioxide, their reflective bodies bouncing radiation back into space. If those two billion seven hundred million people switched to cleaner fuels tomorrow, which is the desired outcome on just about every level, that cooling might no longer counteract the warming under way.

LET THE MARKET DECIDE

The Global Alliance for Clean Cookstoves is trying to tackle the human and atmospheric problems of biomass burning by coordinating the efforts of hundreds of public, private, independent, non-governmental and funding entities[64] that are now partnered under a forty-nation alliance umbrella.[65] In an effort to target its work most effectively, the Alliance selected priority countries, including India, Bangladesh, China, Ghana, Kenya, Nigeria and Uganda. In 2013, it released an assessment for India and declared a huge, if challenging, market potential.

All this spurred private enterprise and free marketeers. Two billion seven hundred million people represent a vast untapped market opportunity. Hundreds of millions of them live in India. Corporate offshoots like the Shell Foundation leapt into the cookstove arena with talk of altruism but also a bottom-line hope that they could profit from a stove with consumer appeal.

Shell Foundation's Breathing Space Program, for example, supported social entrepreneurs, individual companies, and organizations such as Envirofit and Aprovecho Research Center, both based in the United States. Envirofit was launched in 2003 by three University of Colorado students, one of whom had won a Rolex award for inventing a retrofit kit for two-stroke internal combustion engines (like the ones used in auto rickshaws that buzz all over India) that slashed emissions. Cookstoves were the company's next focus. Demonstrating the variety of new funding mechanisms, the company is registered with The Gold Standard for certified carbon credits, collaborates with corporate social responsibility programmes, and teams up with microfinance banks, which have also stepped into the cookstove fray. Micro-loans are often provided through women's self-help groups, a major force in grass-roots dissemination of new products within India.

Harish Anchan, the head of Envirofit in India, met me in the spacious lobby of a Hyatt hotel in the up-and-coming Viman Nagar neighbourhood in Pune, Maharashta, on a Saturday afternoon. Fit and trim, with jeans and a crisp shirt, he'd sandwiched our meeting between a return trip from Nepal, where Envirofit is introducing its stoves, and an afternoon watching the Bollywood hit *Dhoom 3* with his kids.[66]

Before Envirofit, Harish had been working with Videocon Industries, an Indian conglomerate that produces, among other things, consumer electronics, and was looking for a new challenge when he heard about Envirofit's plan to get a for-profit Indian company off the ground. Seven years later, he was still wondering how an electronics engineer got involved with stoves. When he first started, he found a kitchen products distributor and got it to add the Envirofit to their inventory. Then he sat back, waiting for the sales to roll in. They didn't. He realized he was no longer selling televisions.

'It is such a simple technology,' he said, 'but I find it's more related to human emotions and politics to convince the people of the technology.' Harish called cookstoves 'push products', non-profit lingo for something that consumers have to be told they want, rather than something they eagerly pull towards themselves, like cellphones and TVs. How to make a cookstove into a pull product?

'We want to be the Sonys and the Mercedes of these stoves. We don't want to be the Nanos,' he said, referring to the failed 'people's car'.

'Anyone and everyone has an aspiration. You have an aspiration, and you look up the ladder to see, what does she or he have? That is what you need to have.'

Envirofit is now trying every marketing strategy known in India. It uses signage, wall paintings, van campaigns, skits and village-level entrepreneurs to spread the good news of the smokeless stove. Envirofit ran a brilliant television ad in Kannada, in which a boy comes home from school to find his mother coughing while cooking over a smoky wood fire in the kitchen. The boy makes a point of coughing in front of his father, who's in the driveway of their middle-class home washing a big tractor. As soon as he does, both parents come running. 'Daddy! You cannot stand me coughing just once,' the boy says, 'but mother coughs like this daily. Do something about it.' Father and son zip off on a nice motorbike to purchase a smoke-free Envirofit. The closing scene shows the mother serving food to her husband and son. Everyone is smiling.[67]

'That particular ad did wonders,' Harish said. 'Dealers got inquiries. Sales increased up to ten times over.' But television advertising is expensive, and Envirofit was forced to discontinue the campaign after a few months. As soon as it was pulled, sales dropped.

Harish was delighted that competition was cropping up all over India, thanks to companies like Prakti Design Lab, Greenway Grameen, Vikram, Chulika and Serval. The founders of Nirmal, he told me, were Envirofit distributors who became stove makers, mimicking Envirofit's design. Harish had no problem with this.

'It's good to be copied. We need to have multiple cookstoves in the market—ten or twenty Envirofits—to reach 120 million households. We need to have a stove reference and comparison between multiple stoves. We need gold, silver, bronze, so that you can go into any shop, and that salesman will ask you what is your budget. He'll take you to the top, and go from there.'

Harish conceded that government support was needed, not for any single stove but to get people to seek out options generally.

But Harish was wary of too much government. 'I just don't want the cookstove industry to be killed by free subsidies. I can't sell a product if they know it's being given away by some politician,' he said. Chief Minister Jayalalithaa of Tamil Nadu had just announced that she'd be giving away free gas stoves as elections approached. 'So why would you buy a cookstove?' Harish asked.

Whenever Harish is feeling frustrated, he remembers how slow the rise of pressure cookers was. It took ten to fifteen years to sell the first million, he said. 'It's not the Indian way of cooking at all. All these products require habit change.' It's true—you can't even peek at your food while it's cooking in a pressure cooker—yet the whistle of a pressure cooker blowing off its steam as rice and dal and sabzi simmer inside has become one of the quintessential sounds of India.

'We need to be patient,' he said. 'It takes a while for the egg to hatch. In the meanwhile, let's just keep the egg warm.'

BE SCRAPPY

A couple of months later I met Mouhsine Serrar, whose main form of transportation was a motorcycle that he used to get around the red roads of Puducherry (formerly Pondicherry).[68] Mouhsine is a Moroccan-born, American-trained, India-based stove designer and the founder of Prakti Design Labs, which sells improved biomass cookstoves in India, Haiti and Africa. Prakti, an adaptation of the Sanskrit word 'prakṛti', which means 'nature', can be found down Old Auroville Road, with the Bay of Bengal at your back, just past Gaia's Garden. Amid neem and eucalyptus trees, its buildings are constructed with inverted clay bowl ceilings, to conserve cement, and curved brick walls, where pink bougainvillea climb high.

Mouhsine speaks with a French-Moroccan accent and has a faint constellation of freckles speckled across his nose and cheeks. We sit in the office, amid metal Prakti stoves painted glossy red and grey. They come in three main models: wood and charcoal-fueled residential stoves and larger institutional stoves. Prakti stoves start at 1,000 rupees. Some have chimneys. Some don't. The company's residential wood stoves, Mouhsine told me, reduce fuel consumption by 50 per cent and polluting emissions by over 80 per cent. Since 2010, when Prakti was founded, it's sold 5,500 stoves in India. Mouhsine and his team are working hard to scale up operations. They keep two local women on staff to come and use the stove every day, under real conditions, and give the Prakti staff feedback. 'If you give the cleaning lady a broken broom,' he said, 'you will hear it.'

Mouhsine has been living in India since 2008. He travels frequently and packs light. I caught him when he was between homes, searching for a new place near the Prakti office. He only needed it to be cheap and have a terrace. But long before I met him, he was making big money in engineering in the United States. He worked just outside the gates of Los Alamos National Laboratory in New Mexico, a classified site where the atomic bomb was developed, teaching engineers how to use Abaqus, an advanced engineering software used to model, for example, what might happen to a concrete building when large cylindrical objects hit it at high velocities. 'The word ballistics was never mentioned,' he said, smiling. He moved on to crash-testing for BMW in California's Bay Area. 'It was fun,' he said, 'But after three months, it gets to feel a bit empty.'

He had just entered his forties when the second Gulf War began, and he felt the life draining out of him. Still working full-time with BMW, he went to a Green Show in San Francisco, where he was drawn to a tiny booth touting 'Green Career Counseling'. He listed his skills to the lady behind the table, and she sent him some information that mentioned solar cookers.

On a whim, Mouhsine tried using Abaqus to simulate how best to cook a chicken in a solar cooker. He transformed the simulations into real-life solar cookers that he aimed towards the sun from his Berkeley terrace, with minimal success. Then he tried to create a hybrid solar cooker/wood cookstove. 'That was a really stupid idea,' he recalled. He quit BMW and travelled to rural Oregon to learn about wood cookstoves at the Aprovecho Research Center, a rural Oregon non-profit organization that had been developing improved biomass cookstoves for the developing world for over thirty years.

Now hooked on stoves, Mouhsine headed to India in 2005 as the project manager in charge of launching Aprovecho's project there, which was funded by the Shell Foundation. He no longer felt empty. Or bored.

'I've been doing stove work for eight years now,' he told me, practically giddy as he spoke, 'and it's still fun.'

I related to Mouhsine's scrappiness and his idealism. He and I were literally trained at the same school.

When Mouhsine arrived at Aprovecho to learn about stove design in 2005, I had just left the prior year, after nearly a decade of working there.

It was the summer of 1996 when I first arrived at Aprovecho for a three-month internship. Dean Still was the instructor of appropriate technology, which included research and design of fuel-efficient biomass cookstoves. There was also an ascetic man, more spirit than body, who taught organic farming, and a Brit who survived on coffee, alcohol and hand-rolled cigarettes who taught us how to harvest lumber and mushrooms from the woods while cultivating an old-growth forest. I had arrived ready to reinvent myself as a born-again rural girl. I was shaking my suburban New Jersey upbringing and my recent city days in Seattle, relearning skills that everyone had known just a generation or two earlier, basic and fundamental tasks like how to chop wood, bake bread and grow and preserve your own food. Operating on a thin shoestring budget, Aprovecho was an American example of jugaad living.

Aprovecho, which translates from its native Spanish as 'I make best use of', has its roots in cookstoves. It started when a group of American and European friends accidentally invented a world-famous stove while travelling in Latin America. In 1976, while Rajendra Singh was building dams in Rajasthan and the Green Revolution was in full swing in Punjab, the people who would become the founders of Aprovecho were doing earthquake relief work in Guatemala. Someone in the group was a potter and, seeing that people needed stoves, began building them with whatever was available locally. Out of that mud and sand, the Lorena stove was born. A chimney helped get smoke out of the kitchen, and people loved it. Development agencies paid the ragtag group to travel all over the world, teaching about Lorena stoves, but they were still, at heart, more Gandhian villagers than international stove consultants. They eventually purchased the forested patch of land in Oregon where I had lived and set up a non-profit organization devoted to researching the stoves and serving as an example of sustainable living.

With research, they unfortunately discovered that the Lorena stove had a few problems. Yes, it got smoke out of the kitchen, via the chimney, but there was no less smoke. And the beautiful body of the stove, sculpted by hand and offering safety protection from the fire, also absorbed a great deal of the flame's heat. To cook the same amount of food could take twice as much wood.

These facts troubled Larry Winiarski, an engineer and inventor friendly with Aprovecho. He developed a new stove that addressed the failures of the Lorena and incorporated the best available combustion technology. The design principles behind what he dubbed a Rocket Stove include a well-insulated combustion chamber with plenty of room for a natural draft air intake, the use of small pieces of fuel, and a space tall enough between fire and pot to allow for complete combustion. A skirt around the pot directs rising heat along the sides of the pot for added efficiency. Larry's Rocket Stove formed the technological basis for the Envirofit and Prakti stoves, as well as countless others around the world. I didn't realize until years after I left Aprovecho how much overall-clad, devoutly Christian Larry, without a pretentious bone in his rotund body, had helped further global stove design from our backwater in the Oregon woods.

Mouhsine Serrar shared Harish Anchan's desire to create a stove market that could thrive on its own. 'I think what is missing is proper evaluation,' Mouhsine told me as we sat in the Prakti office. 'In the commercial world, evaluation is built-in. If you don't make money, you're out! In stoves, we don't have that. You have USAID spending money, and then they make reports, by USAID, by the consultants, where they say, "We did great!" And they show the reports to the senators and senators approve the next round [of funding]. It has been going like this for thirty years.'

Mouhsine wanted to turn that system upside down. Forget government. Forget the Global Alliance for Clean Cookstoves. 'Get out of the way, MNRE (Ministry of New and Renewable Energy) and consultants! Don't do anything! You cannot drive demand for cookstoves,' he said. 'Only Prakti can drive demands for Prakti stoves. Only Envirofit can drive demand for Envirofit stoves. You don't see signs that say, "Buy Cellphone."'

Mouhsine was tapping into a gestalt best articulated in a 2004 manifesto entitled 'Death of Environmentalism: Global Warming Politics in a Post-Environmental World', written by Ted Nordhaus and Michael Shellenberger, American environmental consultants frustrated by the doomsday language of the left. 'The kind of technological revolution called for by energy experts typically does not occur via regulatory fiat,' they wrote in a later essay in *The New Republic*.[69]

We did not invent the Internet by taxing telegraphs, nor the personal computer by limiting typewriters, they argued. The transition to the petroleum economy did not occur because we taxed, regulated or ran out of whale oil. Those revolutions happened because alternatives appeared that were vastly superior to what they replaced.

Mouhsine was animated by the same spirit as Nordhaus and Shellenberger. Granted, he might have been ignoring the very real push/pull chasm between stoves and cellphones, but what he was not ignoring was the crores of rupees spent on creating reports about cookstoves when that money could have gone directly to help develop better cookstoves themselves.

The Washington DC–based Global Alliance for Clean Cookstoves does fund stove projects directly but it regularly seeks out professional advice to help it decide where to go with its hugely ambitious programme. These consultants, Mouhsine said, 'lack the enterprise spirit'.

Prakti Design Lab has that spirit in spades. Initially funded with Mouhsine's savings from his engineering days, it is beginning to receive outside funding, from the Alliance and a recent innovation award, but Prakti is also depending on the social entrepreneurship model, from viral videos to volunteers, to help get stoves to people.

Governments and non-profit organizations have 'received millions of dollars to do marketing', Mouhsine continued. 'It's a frickin' waste of money. It's not working.' He wasn't advocating for freebies or even grants, but a different sort of financial support. 'Just take ten companies and work with them like GVEP,' he said, speaking of Prakti's contract in the Democratic Republic of Congo with Global Village Energy Project, an international non-profit organization that allocates funding on a contractual basis. 'Look at their business plan, with deliverables, and give the money in a metered, responsible way. It's not a grant. It is a contract. We have to deliver to receive that money. We are not an NGO asking for money.' Yet the market for cookstoves is still emerging. Conventional loans are hard to come by, grants can disappear into a void, and angel investors are few and far between.

Mouhsine and I left the office to head out back, where an open-air lab loaded with testing equipment stood next to a shed where a woman fed fuel into four wood- and charcoal-burning stoves. 'This is the death camp for stoves,' Mouhsine told me. Prakti puts its stoves under stress to find out how long they can last before they burn out. Most of Prakti's stoves are manufactured in a factory in Chennai run by Mouhsine's partner Anandan Sundarmurthy whom Mouhsine described as 'The best. Manufacturer. Ever.'

We stood by a large workbench with an emissions hood dangling above. 'The stove companies need to be like Apple,' Mouhsine said. 'Not in like being fancy, but in making a market that is totally new, where you have to build everything, where nothing is proven, where you have to take risks. You have to have one entity that has the control from the production to the distribution,' he continued, 'because it is difficult and you need to try different things and you need to try them very fast. If I want to change the design, I can tell Anandan over the phone or send him a sketch by email. In the three hours I'm driving from Puducherry to Chennai, he's doing the CAD, and by the time we finish dinner, we can be holding the stove.'

One of the fires in the shed began smoking, and the woman made some adjustments to the wood. The smoke subsided.

'The same way you have to innovate to make a better stove cheaper, you have to innovate on how to make distribution and marketing more effective,' said Mouhsine. 'You have to realize that a stove is not a TV.'

Mouhsine too stressed the need to break away from the poverty angle. 'Our tagline is *smart stoves for smart cooks*. It's not about the poor dispirited woman, we feel so bad for you, here is something that will help you. No, you buy it because you're smart.' He told me about a customer in Nepal who told one of Prakti's female representatives that the stove gave her confidence. Even Mouhsine thought this was a stretch, figuring something was lost in translation. But the woman explained that she was a day labourer who had to go out in search of work every morning, negotiating anew her rate for each job. 'When you have smoke in your house, you smell like a chulha,' she said. 'You smell poor.' She no longer smelled like woodsmoke, and the absence of that poverty signal allowed her to negotiate a better rate. 'You're not making things for poor people,' Mouhsine said, 'you're making things for people.'

Mouhsine conceded that the Alliance was raising awareness, but added that 'it's almost a waste. It's like the things that we need five years from now they are doing now. Why organize a big conference, meet the governments, the ministry, to tell them about stoves if there are no stoves available for distribution, if there is no product for that area, if there is no marketing? There are more testing centres than stove companies. What are they going to test? And when they test stoves, they find that the stoves are bad.'

We headed back to the office. 'I'm committed to not be part of the problem,' Mouhsine told me, determined. 'The problem is that when all these people go to the Global Alliance conferences, everyone says everything is great, because if you say it's bad, you won't get funding. I am committed to my work, I will not do that.' Many people I spoke with slipped repeatedly off the record, dancing around the delicate, coy about the inefficiencies, the corruptions, the misguided efforts that didn't pan out. Mouhsine's words were refreshing.

'They will hear what I am thinking,' he promised. 'Now, for the community, people should stop being polite, should call out the crap, the waste. As they say, "Insh'allah, we can do it." A lot of people need to be fired, from USAID to the Global Alliance, just so that the next person doesn't do the same mistakes. But right now, everybody says, "It's nice, it's nice." The stove community now is almost like a dynasty. You go to the conference and it's the same freakin' people, the same speech.' We settled back in the office.

'The question is how to help poor people,' he said, 'how to do development.' Even with all the infrastructure existing around an entity like the Global Alliance, the stoves were still not being adopted by people. 'Consultants make money, everybody makes money, but the problem does not get solved. Why does it not get solved? We don't even get to that point. You pour money and after five years, ten years, it doesn't work, and you just move on.'

Mouhsine looked at me and shook his head. 'Everybody makes money, except the poor.'

UP IN SMOKE

As stove designers continue to scramble to perfect a clean cookstove, there are still unresolved questions about what, exactly, a clean cookstove is. Gautam Yadama said, '"Clean" is a nebulous term.' 'What are the metrics?' he asked me when I spoke to him over the phone.[70] 'Who is calling them improved and are they improved?'

The International Organization for Standardization (ISO), a worldwide federation of national standards bodies formed in London after World War II, is trying to address this issue. ISO has systematized units of time, measurement and physical objects from freight containers to infant car seats. In 2013, it established a clean cookstove technical committee aimed at standardizing stove testing. The task won't be easy. An improved stove might be less harmful for individual humans but still spew devastating particulates into the atmosphere, or be climatically 'clean' and terrible for the cook's health. It all goes back to the non-ideal world inside a combustion chamber—to those kinetic and thermodynamic constraints. Fuels vary. Moisture varies. Humans vary. As one young designer told me, 'You start 100 fires and in the first ten minutes you get 100 different burns.'

'Thirty years of research,' Kalpana Balakrishnan said, 'has really not produced a cost-effective way of burning wood.'

Yet biomass cookstove companies blaze forward. In 2012, organizations partnered through the Alliance claimed distribution of 8.2 million clean cookstoves in a single year.

Envirofit has sold 3,75,000 stoves since it started.[71] And the Indian government says it distributed an impressive thirty-four million stoves from 1983 to 2002 as part of its National Programme for Improved Cookstoves (NPIC).[72] In 2009, the Government of India launched a new programme, the National Biomass Cookstoves Initiative (NBCI).

But distribution is just one step in the path away from smoky fires, as I saw in Nandal. Continued use of those stoves is something altogether different. A few weeks after I met Kirk Smith in New Delhi, he returned to the villages in Gujarat where he first worked in the early 1980s.[73]

'I'm afraid that I'm going to see that nothing has changed,' he'd said ruefully in the Hauz Khas cafe before he went. Instead, he told me later, he saw vast changes to the place he once knew. People chatted away on cellphones, and most houses had electricity, satellite dishes and running water. But one thing hadn't changed: most homes still used chulhas for at least some of their cooking. Development, he told me later, has become disconnected from cooking.[74]

The Jameel Poverty Action Lab (J-PAL), established by the Department of Economics at Massachusetts Institute of Technology (MIT) in 2003, collaborates with researchers all over the globe in randomized controlled trials meant to shed light on some of humanity's most debated and intractable social ills. J-PAL has studied school testing methods in Madagascar, rainwater harvesting in Brazil and numerous projects in India.

Often, the studies help illuminate what works, but just as revealing is discovering what doesn't. Rema Hanna, associate professor of public policy at the Harvard Kennedy School, collaborated with MIT economists Michael Greenstone and Esther Duflo (who is also the J-PAL director) to see how much an improved cookstove improved user health in Odisha, India. But just four years after the stoves were first distributed, the researchers could find none in existence. No stoves meant no improved health. 'We find no evidence of improvements in lung functioning or health, and there is no change in fuel consumption (and presumably greenhouse gas emissions),' they wrote in the resulting report, called 'Up in Smoke'.[75]

'That was really damaging,' Kirk Smith said about this report. 'Those stoves were bad stoves. Of course people stopped using them. We did eight years of feasibility studies before we introduced a stove.' He was referring to his five years of oft-cited research in Guatemala known as RESPIRE—Randomized Exposure Study of Pollution Indoors and Respiratory Effects—a long-term study of children's health in homes with wood cookstoves.

'One thing that people take away from this study is that stoves don't work, but that is wrong,' Rema Hanna told me over the phone.[76] 'The larger-scale distribution has been for cheaper stoves. Sounds great on paper, but they break, people don't like them, and they're not effective in practice even if they do well in a lab.' Hanna wanted the techies to join forces with behavioural scientists. 'We need to bring together all the different fields to create a stove that works and that people will use,' she said. 'We need to find something that is human-proof!'

After visiting Nandal and also finding what the 'Up In Smoke' researchers had found, I had sought out Dr Priyadarshini Karve.[77] She had left ARTI to form Samuchit, a private company that sold improved cookstoves developed by ARTI. The Samuchit office was a few flights above a side street just off leafy Law College Road in Pune. Priyadarshini was in her early forties, snug in a black fleece. She sat at a desk in an open office that doubled as a storage room, with stacks of stoves and boxes spilling from the corners. She had worked with cookstoves since she first developed one as an undergraduate. Samuchit made adorable biomass stoves that appealed to the middle class: to those who cared to reduce their carbon footprint or to the Brahmin followers of Guru Raghavendra Swami from Karnataka, who eschewed modern amenities such as LPG and pressure cookers. Priyadarshini discovered that when sales went up in the middle class, they also rose among lower classes aspiring to achieve the lifestyles of their better-off family members or neighbours.

As for finding a human-proof stove, Priyadarshini agreed that there was a disconnect between stove work in the lab and the desires of the women who cook.

'A very important aspect which has been ignored by everybody in this whole discussion of emissions and health impacts and climate change impacts is that the stoves have been designed keeping in mind all these needs. Somewhere in this process,' she said, 'they have lost their ability to cook.'

She was thoughtful about the dilemma, and contrite about the role she and other designers played creating stoves no woman would want. 'I think it is a bit of dishonesty on our part as stove manufacturers to claim that, just by selling so many stoves, we have solved the problem,' she said. 'We haven't, because those stoves are not the exclusive stoves that are in use.'

But could that actually be a good thing? I found one answer to that question in Tamil Nadu, among Christians and magnificent granite boulders. But on my way, I needed to stop for a cup of coffee in Bengalaru, the capital of Karnataka, for a conversation with another former ARTI employee, who doesn't think marketing to the middle is the answer. The answer might not be cookstoves at all.

Warm coffee did little to take the chill out of the air in Bengalaru, where I met with Karabi Dutta, a woman who worked at ARTI from 2000 until 2010 when she left to become an independent household energy consultant.[78] The outdoor area of Café Coffee Day, across the street from the gleaming new MG Road metro station, was filled with young couples smoking cigarettes in the blustery air and a gaggle of boys who were all Bollywood, sporting sunglasses on an overcast day.

Karabi had been with ARTI in the final years of the NPIC, the government's earlier stove programme, when it was tasked with monitoring and evaluation. After almost twenty years of distributing millions of stoves, the government finally tested them and discovered they didn't work. When that programme ended in 2002, the national government passed responsibility for clean cookstoves on to the states, and most failed to take up the initiative. Of the thirty-four million stoves that the government had supposedly distributed, virtually none could be found in working condition. The programme was criticized for its poor stove design, high cost, low uptake rates and a heavily subsidized system that defeated any nascent consumer stove market. It came to symbolize everything that could go wrong with a cookstove programme.[79]

Karabi was most concerned about the lack of a consumer market. For many in India, there's no place where a customer can look at, say, ten different stoves and decide which one to purchase. It's impossible to shop for a cookstove in the way someone might shop for a refrigerator or a television set, browsing a range of options from basic to deluxe. There is little consumer choice, because the market is missing.

'We are deciding which stove is good for them,' said Karabi. 'But you cannot decide for me.' She also pointed out that markets can be distorted or destroyed by counterproductive generosity: 'If it's free, they take whatever it is you give them.'

Good stoves have come to the market. At one point, there was a lot of buzz about the Philips stove, a forced-draft model that used a battery-operated fan and outperformed everything else in the market. It retailed for under Rs 5,000 in 2006.

Karabi said that Philips had been ready to go into production, hoping to market their stainless stove during Diwali, the Hindu fall festival of lights when Indians rival American Christmas shoppers in their spending. Philips just needed a tax break to bring the price down to a point Indians could afford. But the government refused, because the stove didn't target the neediest class, Gandhi's humble villagers. Philips moved its business to Africa, where it got more support.

Meanwhile, stoves sponsored by government programmes in India were either undesirable or inefficient. 'They were very bad losers, the people who ran the project,' Karabi said. 'They did not believe, until even in 2010, that the stoves they had disseminated were not really efficient enough.'

I asked her if the problem of clean cookstoves was beyond solving.

'It is not a problem which can be solved very easily,' she answered. 'No one solution will work. You have to allow stove stacking.'

OPTIONS

In some places, development is helping to make the question of improved cookstoves moot and hinting at a future without open fires. Even as they pursued the perfect biomass stove, many developers acknowledged to me—almost as an aside—the reality of stove stacking, in which households use multiple cooking sources. It's exactly what happens in middle- and upper-class homes, where we seamlessly segue between gas ranges, microwave ovens and electric kettles.

One vision of what stove stacking can look like is evident in the southern state of Tamil Nadu, one of India's most developed areas, where gneiss and granite outcrops run like an ancient current below and above the surface. These rocks are the defining feature of the landscape south-east of Bengaluru, from where I was travelling out, rising like islands from rice paddy seas. I was on my way to Krishnagiri district, where Envirofit had been active distributing stoves on a wide scale through the non-profit Integrated Village Development Programme (IVDP).

What I found around the compact city of Krishnagiri could be the future of India. At nine in the morning, much of that future was headed to school. The streets were jammed with kids of all ages, their coloured uniforms announcing allegiances to school, village and town. Daughters sat pillion on their fathers' motorcycles. Boys walked hand in hand. Girls in blue salwars with yellow ribbons at the base of each twin-looped braid rode their bicycles, dupattas neatly folded and pinned in back.

Fields were thick with banana trees and mango orchards, and the iridescent tail feathers of blue rollers occasionally streaked the sky. A stone Hanuman sat sheltered in the middle of a bhindi crop. Shiva's trident and Jesus' cross gestured skywards.

In a village called Elathagiri, I climbed a narrow flight of stairs that took me above the Ladies and Gents Tailors shop and into the home of a woman named Emily Teresa. It was a basic four-room apartment painted in bright hues. Though living on about Rs 8,000 per month—the income her husband earned as a weaver—Emily Teresa owned a new refrigerator, an Aquaguard water filter, a blender and a television set. She also had a stack of stoves. A pressure cooker whistled atop the two-burner gas stove on her kitchen counter, releasing an aromatic hint of dal into the air. It was her favourite, and its LPG cylinder could last almost forty days between subsidized government rations. She also used a one-burner kerosene stove that was tucked below the counter to heat water, but the two litres of monthly subsidized kerosene ran out in just a couple of days. So most of the water heating happened on Emily's third stove, which was back down the stairs, along the side of the house: a three-stone fire.

Back upstairs, her fourth cooking device was an Envirofit, which she could fuel with wood or the coconut fronds that littered the edges of the cornfield behind her house. Emily lifted this cookstove from its spot in a corner and carried it outside to the terrace to show it off in the daylight. The size of a two-gallon bucket, this small portable metal stove was once orange but the front of Emily's was blackened from use. On either side were handles for easy transport and at its base was the mouth of the firebox, with a metal grate to keep the wood elevated and oxygenated, sticking out like a tongue.

The Envirofit was one of the 22,000 that IVDP had distributed in the district. Emily, along with 1,75,000 others in Tamil Nadu's Krishnagiri district, belonged to one of the thousands of women's microfinance groups this non-profit organization had organized. These groups helped transform the household economies of women like Emily, who was forty-five years old and cared for three children. The stove hadn't been a giveaway. Emily had purchased it, like the other modern devices in her home, through rotating loans within a group of twenty women. Each month she contributed a small amount of money to a shared pool, and when she needed funds—whether for a cookstove or to put her kids through school—she withdrew what she needed.

Emily used the Envirofit frequently in the first month, she said. But a year after she'd got it she was using it only a few times per week, or per month. The novelty had worn off and the stove had burnt out despite being under warranty for five years. She didn't know how to go about getting it fixed. She pointed at the rust and the holes in the metal of the combustion chamber.

Andhoniammal, Emily's mother-in-law, lived next door and was also dissatisfied with the improved stove. 'You need one person to stand with the Envirofit stove and manage it,' she said. 'That person has to keep breaking the wood sticks into small pieces and putting it in,' she continued. 'In the chulha, we can put a coconut frond and leave it there and the water will heat up. In this Envirofit stove, the work doesn't get done quickly.' On a later visit to her home, I asked to see her Envirofit. She pointed to an inaccessible storage space above a doorway.

Emily's sister-in-law Mary Stella had a sweet smile and a zigzag gold earring that snuck up her earlobe. She lived down the road from Emily and sold milk from a few cows she kept. Her husband was a tractor mechanic. Mary Stella had all the stoves that Emily did, as well as a single-burner electric induction stove, which looks like an ordinary electric hot plate but is a completely different technology. Induction stoves use electromagnetic induction to heat a compatible ferromagnetic metal pot and are proving increasingly popular in India and in the West. Since the heat travels directly to the food, and no heat is emitted once the pot is removed, induction cookers are safer and much more efficient than other electrical or gas stoves.

As for Mary Stella's Envirofit, she also admitted to hardly using it. All three women agreed that the Envirofit was inconvenient. It just didn't fit into the design of their lives. And yet they each had bought one. I asked them why.

'First one person bought the stove,' Mary Stella explained. 'She told us that she kept the stove inside the house itself and made biryani. There was very little smoke. So, we felt like getting one too. So, we bought it of our own accord. I've used it only four or five times in the year, but I still don't regret buying it especially because we are only paying a monthly instalment of 100 rupees.'

Peer pressure can be a powerful motivator.

Atul Gawande, writing in *The New Yorker* about push products like improved cookstoves, highlighted the need for face-to-face contact.

'In the era of the iPhone, Facebook and Twitter,' Gawande wrote, 'we've become enamored of ideas that spread as effortlessly as ether. We want frictionless, "turnkey" solutions to the major difficulties of the world—hunger, disease, poverty. We prefer instructional videos to teachers, drones to troops, incentives to institutions. People and institutions can feel messy and anachronistic. They introduce, as the engineers put it, uncontrolled variability.'[80]

But humans are social creatures. Gawande quotes communications scholar Everett Rogers: 'Diffusion is essentially a social process through which people talking to people spread an innovation.'

So why didn't the Envirofit's failure among the women in Elathagiri feel to me like the failure it had been in Nandal? There, a broken stove meant a return to a smoking chulha. Here, there were other options, including traditional, LPG and electric induction cookstoves, or a combination of all of the above.

As I left Elathagiri, I passed by dosa-wallahs feeding thick stalks of coconut fronds and six-inch logs into their streetside stoves as they ladled dosa batter onto the griddle and filled the dosas with potato masala. Next door were freshly slaughtered chickens, plucked and gutted and suspended in a row, their yellow feet startled into stars. It was the end of the day, and the striking rock formations perched on hillsides beside the road had transformed into silhouettes, looming dark against a brilliant fuchsia sky as the sun disappeared. Some of these formations appeared to be eternal and unchangeable, pushing the bounds of gravity, teetering, anticipatory. Others sat roadside, dwarfing my car and split asunder, as though they had tumbled down just an hour or two before I rounded the bend. Things that appear rock-solid can cleave when no one is expecting it, as soon as a tipping point is reached. Something immutable can transform.

LEAP

I had first met Emily Teresa along with dozens of other Elathagiri women who were members of two local self-help groups. We'd gathered at the steps of the Our Lady of Refuge Church, a grand white building with screaming turquoise blue trim. In that cluster of women, I saw one possible future for India, revealed in that bouldered region of the south. Ten years earlier, all those women had used chulhas and nothing but chulhas. By the time I arrived, they were debating the merits of LPG versus kerosene versus Envirofits and more. They had choices, the result of multiple policy decisions that had helped Tamil Nadu, by 2015, to hold the leading position in India on eight out of nine development parameters such as economy, power, roads and health.[81] The state has invested not just in economic development, but in human development, ensuring universal education—especially for marginalized Tamilians through reservations—which raised the literacy rate and brought more citizens into the growing economy in a positive feedback loop. The state was also the first in India to introduce the free midday meal scheme. The Tamil Nadu model was one in which growth was accompanied with social justice and equity.[82]

When I asked the women which stove they preferred, they did not hesitate in their answer: 'Gas is the best and quickest.' In fact, only one woman dissented, an elder who still preferred the taste of spices ground on a stone and food tinged with woodsmoke. The other women were willing to sacrifice a little taste for convenience and speed. If only gas rations weren't so limited and availability wasn't so meagre.

Given the fact that modern fuels are what these chefs want, and that women worldwide have refused to adopt, on any meaningful scale, the 'new and improved' cookstoves that have been pushed on them, many researchers are beginning to question whether it's wise to maintain their dogged pursuit of the perfect biomass cookstove. What if all those efforts went instead to finding ways to expand access to the proven technologies already cooking up food in the developed kitchens of the world?

Priyadarshini Karve of Samuchit had suggested as much in Pune. 'There is a certain aspirational aspect which no one is really paying much attention to. That is a huge part of the story. People want LPG. Period,' she said. 'But then again, if it has to be a wood-burning stove, people seem to want something that looks the same as what they are using,' the tried-and-true chulha. She admitted this was a bit illogical, but that's how it was.

'Whenever I'm in the field, I ask women what they want,' Kirk Smith told me in the New Delhi cafe. 'Maybe I lead them a little bit,' he said with a smile. 'I ask, "Would you rather cook standing up … like they do in the movies?" They always say yes. But then they might say, "But my mother-in-law wouldn't like that."'

Back in a house in Nandal, I spoke with a woman as her daughter-in-law silently made bhakris over a traditional chulha. A Bharat Laxmi improved stove lay abandoned in a corner. 'She has asthma,' the elder woman said, gesturing at the younger woman, 'and it gets worsened by the fire.' An LPG connection was on order, she added, but the health of the younger woman was not the reason they were switching to gas. 'Everyone else has one,' said the mother-in-law, almost defiant. 'Why not us?'

Why not? At times, change in India seems exasperatingly slow, but change happens. 'One generation abandons the enterprises of another like stranded vessels,' wrote American transcendentalist Henry David Thoreau.

'People say it'll never happen in India,' Kirk said, 'but it happened in China. It happened in Mexico and the rest of Central America. Quickly.'

'I'm challenging the assumption that the poor can't afford clean cookstoves,' Kirk said, as we sat in the Hauz Khas cafe. In 2000, Indians took nearly two billion sick days attributable to indoor air pollution.[83] Sick mothers, sick fathers, parents and grandparents tending to sick children. Lost school days. Lost pay. Lost lives.

'LPG prices can vary. I know this,' Kirk said, leaning forward. 'But if you calculate the social benefits of health, LPG is cheaper.' As a society, we don't think twice about subsidizing vaccinations, he argued. We do that as a routine part of keeping healthy, as a social investment. It should be the same for clean cookstoves.

If India really wanted the improved biomass cookstove market to take off it would need to reduce or eliminate gas subsidies and make LPG prices reflect their true cost accounting. That would make gas too expensive for the millions who can barely afford it now, but the Envirofits and Praktis of the world would stand a better chance if they were operating in a free and level market. However, the cleanest cookstoves will likely always be the most expensive, and taking this tack could mean more people end up cooking on old-fashioned biomass stoves, not less.

Another alternative is for India to make government subsidies of gas so generous that this cleaner form of energy becomes available even to the poor. Today, red LPG cylinders are transported precariously through fractured supply systems, tethered to the sides of bicycles, loaded up on lorries, and lashed to the backs of motorcycles. And when they arrive at their destinations, their sticker prices are beyond the means of all but the middle class.

In 2013, Bjorn Lomborg, a former director of the Environmental Assessment Institute in Copenhagen, wrote a *New York Times* op-ed arguing that the poor need access to cheap fossil fuel for their own health and for everyone who stands to benefit from a healthier global climate.[84] It seems counter-intuitive, that getting more people to use fossil fuel instead of renewable biomass could be good for the environment, but given all the new data on how horribly polluting burning biomass is, that seems to be true. Lomborg's view echoed that of Kirk Smith, who wrote a piece in *Science* more than a decade earlier, entitled, questioningly, 'In Praise of Petroleum?'[85]

Kirk's piece disputed the fact that biomass is truly sustainable given its human and climatic health impacts. And, even if renewable, the article questioned whether biomass can be harvested at sustainable levels. Kirk argued that switching domestic households to fossil fuels would add only 2 per cent to overall global greenhouse gas emissions. Regarding the supply side, he argued that if modern fuels were available for industry—a far larger consumer than households—they should be available for domestic use. For everyone.

Finally, he questioned the alternatives that have been offered to the poorest half of the world, to the Seemas and Vanitas who receive cheap stoves that don't work and the millions of unnamed women who bear—on their backs and in their lungs—the brutal effects of biomass burning.

'Does it make sense to ask the poor to take on novel devices and fuels that have never been tried elsewhere?' Kirk asked. 'Shouldn't it be those that produce the most greenhouse gases and have the resources and technology to do something about it who should shoulder the burden of testing and using new low-greenhouse-gas, high-efficiency technologies?'

Kirk is frustrated, a prophet who has crunched the numbers, published them widely, spoken to anyone who asked, and seen his life's work enfeebled by policy stagnation and cultural limitations.

'We can't push climate change on to the poor,' he said. 'If the Indian fleet of vehicles had a half a per cent increase in efficiency, it would offset the conversion of every biomass stove to fossil fuels. So don't tell me that cookstoves are the problem.'

The IPCC estimates that achieving universal access to electricity and clean fuels would collectively cost between Rs 4 lakh crore and Rs 6 lakh crore per year until 2030. Effects on greenhouse gases would be minimal.[86]

Supporting Bjorn Lomborg and Kirk Smith's position was the epidemiologist Kalpana, who frequently collaborates with Kirk. She too had seen the limitations of improved cookstoves. She had analyzed the blackened filters of air monitors that showed that 'clean' cookstoves were not much of an improvement over the traditional chulhas. She made a moral argument against improved cookstoves.

'Are you justified in saying that it's okay to be just a little bit better?' she asked, questioning the promotion of stoves that are known to still be lethal compared to gas. 'We don't have the ethical ground for arguing for all the intermediates,' she said. 'That's the problem.'

Complicated questions remain. How can India get LPG to more people when it is already importing nearly half of its fifteen million tonne annual consumption, much of it from the volatile Middle East? Government subsidies are already stretched thin, and there is increasing pressure from the global community to lessen human dependence on fossil fuels.[87] Still, 'the only thing we know that's ever worked is gas and electric', Kirk said bluntly.

Karabi Dutta, the stove consultant in Bengalaru, said the same thing, though she leaned towards electric stoves, more specifically electric induction cookers. 'They are fuel-saving. They are clean. If I look at it from the health point of view, this is something that is going to solve a lot of problems. Even more than gas.'

Gautam Yadama also agreed that investing in electrical cookstoves made more sense than trying to dream up the ideal wood cookstove. 'We might have to think very seriously about leapfrogging,' he said, 'and not tinkering at the margins with improved biomass stoves.'

Even the Global Alliance of Clean Cookstoves acknowledges the advantages of ultimately abandoning biomass stoves. 'If people can afford to and are able to access the cleaner cooking technologies, including electric and LPG stoves, that's wonderful from our perspective,' said director of programmes Sumi Mehta.[88] 'But we also know that in the short term not everybody's going to be able to leapfrog to that.' Of the three billion people burning biomass, at least a third have little hope of moving to gas or electricity any time soon. For them, she says, the Alliance will continue to invest in creating a cleaner biomass stove, no matter how challenging the job. The stoves will have to be, in Gautam Yadama's words, 'good enough'.

The millions of Indian homes that most desperately need to transition away from dangerous cookstoves are typically the same homes that are also in want of basic electricity. Even as access to gas slowly expands, focusing on getting electricity to the six hundred million Indians who lack it could solve the cooking quandary and illuminate the poorest corners of the country at the same time. The question is how to expand the electric grid without building more carbon-spewing coal-fired power plants or risky nuclear plants along the shores of tsunami-prone South Asia.

The answer comes mostly from above ground, from solar, wind, biogas, microhydel. And the answer is not far away but close at hand: localized, small-scale power plants. Energy seekers globally are increasingly turning to these renewable energies. In the developing world, decentralized micro-grids provide power to rural communities immediately and can also be designed to link up with slowly expanding grids if and when they come.

Costs for simple renewable power systems and for smart-grid technology (based on automated information that helps conserve energy) have both dropped substantially, making the transition possible like never before, especially in places currently relying on expensive and polluting diesel generators for lack of other energy options. Prices for renewables have dropped so much since 2009 that, in some markets, producing power from renewables is now cheaper than from coal or natural gas. But subsidies play a huge role in this price point, meaning nations have a choice of which version of power production they want to incentivize. According to the International Energy Agency, fossil fuel subsidies were over four times the value of subsidies to renewable energy in 2013, leaving much room to encourage renewables over oil and gas.[89]

The conversion away from non-renewables has been rapid, with increasing numbers of countries making larger commitments to renewable sources. Coal and oil still dominate global power production, of course, but wind and solar accounted for more than a quarter of the growth in electricity generation in 2013.[90] As with organic agriculture, this is still a tiny fraction of the whole. But here too we could be witnessing a seed ready to sprout. Domestic renewables should also prove more alluring as the Middle East continues to fracture in violent upheaval. If global oil prices go low, however, the pitch will be more difficult.

Micro-grids are already popping up across India, in Maharashtra, Karnataka, Uttar Pradesh and beyond. In Tamkuha, even a simple solar photovoltaic system was beyond its inhabitants' economic reach, but the Bihari village was swimming in rice husks. Savvy designers transformed the waste into biofuel with a gasifier, and the villagers strung power lines through the village, carrying enough power to bring basic electricity to hundreds of homes. The concept has been replicated in eighty more micro-power plants serving India as well as countries in Africa.[91]

Because a majority of Indians depend on agriculture for their livelihood, and the waste from field and farm animals is plentiful, research on more efficient power production from these sources should pay off. Sunil Dhingra, a senior fellow at The Energy Resources Institute (TERI), a New Delhi-based policy centre, estimated that India produces six hundred million tonnes of agrowaste each year, nearly a third of which is not used. This is a potentially huge resource of fuel lying in wait.[92]

The scale of renewable energy projects will continue to be debated, especially as some of the largest wind and solar projects in the world are under consideration within India's borders. These megaprojects are best approached with caution; like large dams, the massive ecological and social impacts of these schemes can outweigh their benefits. The 2014 national budget included funding for large solar power projects in just four of India's twenty-nine states. The Centre for Science and Environment (CSE) observed that this money might have been better spent setting up 20,000 mini solar plants that together could provide power to as many as fifty million rural households.[93]

Just as reducing food waste eases the pressure on how much food needs to be grown, energy conservation can also pay for itself. Fossil fuel subsidies were four times the amount invested globally to improve energy efficiency in 2013, a wasted opportunity to reduce wasted wattage.[94]

Just as Karabi Dutta had hinted at, electrical induction stoves like the one in Mary Stella's kitchen hold huge promise. Their use is beginning to take off in India. They are incredibly efficient, cost about the same as mid-range improved biomass stove, and if run on micro-grids powered by renewables, they could provide a means for millions to move away from biomass altogether. All these efforts seem to be a more hopeful answer for how to make places like Nandal village truly smoke-free. Perhaps the bhakri bread won't taste the same, but urgent climate and development needs could be addressed and light would be safely and reliably available for the night-time work of both parents and children. Expanded access to electricity could also liberate women from the time-consuming task of fuel collection, keep household air free from smoke, and prevent black carbon from clouding up our atmosphere and melting the world's ice reserves. Given such benefits, those delicious singed spots on the bread seem to become less and less important.

Early in 2014, the Anil Agarwal Dialogue gathered people working on energy access and renewable energy to New Delhi. The room at the India Habitat Centre was packed, and I saw some familiar faces; Priyadarshini Karve had come from Pune and Mouhsine Serrar from Puducherry. During the conference, which was coordinated by CSE, there was talk of cookstoves, but many times it was in the context of moving beyond them.

CSE's Anumita Roychowdhury spoke of the 'chulha trap', in which the number of traditional cookstoves in use stubbornly refused to go down. 'Go to a film and you see the government warnings,' she said, referring to the large 'SMOKING CAUSES CANCER' that screams across the screen every single time an actor takes a puff, 'but our rural people are surrounded by the equivalent of four hundred cigarettes a day.'

'Are we trying to sell them something that *we* are not willing to use?' asked Dr Priyadarshini Karve to the hundreds gathered.

There was consensus that the best use of biomass was within biogas digesters or gasifiers, which produced more efficient power than could be produced in a cookstove, whether traditional or improved.

Micro-grids and renewables dominated the dialogue. 'Mini

grid—or decentralized generation of power—offers exciting possibilities of reducing India's energy poverty,' said Chandra Bhushan, deputy director general of CSE.

'I don't see one reason why decentralized solar power and decentralized wind power should not get mainstreamed in the next few years,' said Ajay Shankar of the National Manufacturing Competitiveness Council, a body set up by the Indian government to encourage manufacturing.

Dr Satish Balram Agnihotri, whose last name, in the time of the ancient Vedas, would have revealed him to be a fire priest, was the newly appointed secretary of the Ministry of New and Renewable Energy (MNRE). He saw the solution to India's energy problem perching on three legs.

'We must have renewable energy as a source, energy efficiency as the way of utilizing it, and a smart grid wherever we are setting it up,' he said. 'This is the trinity.'

Nowhere was there mention of clean biomass cookstoves in Agnihotri's trinity. Seema and Vanita's lives seemed far from the room, but Teresa and Mary Stella's were closer. Sometimes the only way forward is to let go of the past. The fire god Agni has served humanity well for millions of years, cooking the provisions that sustained us for untold generations and that sustain nearly three billion living today. But Agni has also been unforgiving, on human bodies, on atmospheric chemistry.

It may be time to extinguish the match. Cooking, that quotidian task, relentlessly necessary, done overwhelmingly by women, could be so much cleaner. To commit to leaving smoky stoves behind could begin to answer India's energy and health crises, as well as the planet's climate change crisis. The transformation will start at Seema's hearth in the west, at Emily's stoves in the south, and end when the lungs of their children are pink and the skies over South Asia return to azure.

4

VAYU

AIR

With rhythmic beats alternately covering and uncovering parts of the sky; scattering the clouds and thereby exposing flashes of lightning to view; and breaking down the rocky cliffs of the mountain ranges into pebbles and powder, the mighty vibrations of the vast wings of Jatayu proclaim his arrival from a great distance.[1]

—*Mahaviracharita,* Exploits of a Great Hero

VANISHING VULTURES

'*Ab gidh bhee kum ho rahe hain?!*' Ram Gopal asked his colleague Asad Rahmani at the end of 1996, waving a newspaper clipping in his hand. 'Where have the vultures gone?!'[2] The two worked together at the Centre of Wildlife, at Aligarh Muslim University in Uttar Pradesh, where Rahmani was chairman and Gopal worked as an animal keeper. Vultures were ever present over the South Asian landscape and then they were suddenly and inexplicably gone.

Very few Indians noticed the empty air overhead. Vultures were rendered invisible by their commonness. Perhaps something in us didn't want to see them. They are cross-culturally uncharismatic, with their featherless gray heads, pronounced brows that make for permanent scowls and oversized blunt beaks that are sometimes capable of splintering bones. The broad wings of South Asian species can reach up to eight feet tip-to-tip, casting a great shadow from above as they circle, drawn by the distant sight of the carrion upon which they feed.[3] On the ground, they are as massive as they are clumsy, lurking along the banks of the Ganga and hopping precariously on to dead bodies. The world over, these voracious scavengers, their naked faces seen buried in rotting flesh along the roadside, are viewed with disgust and fear, and associated, rightly, with death. When disturbed or frightened, their main defence is to vomit. Instinctually, we look away.

Just twenty years ago, India's vultures were some of the most abundant species of bird on earth. There were at least forty million flying across the skies of India, Nepal and Pakistan, casually disregarding contested and shifting borders below. For a subcontinent where religious and cultural mores restrict the handling of the dead, human and animal alike, vultures were unsung servants, a natural and efficient disposal system. They scouted the countryside, searching for the remains of goat carcasses that Muslims might not eat if they hadn't been killed in a halal manner and cows that many Hindus would eschew. Vultures soared over the cities, cleaning up roadkill and picking at the detritus left behind by India's ever-expanding populace. In New Delhi alone there were at least one hundred thousand vultures. In Mumbai, they patrolled the Towers of Silence where Parsis, a small but ancient religious group, laid out their deceased for the vultures to consume in a ritual known as a 'sky burial.'

Photographs in the archives of the Bombay Natural History Society (BNHS) document the bountiful days. One photo from the 1980s depicts 6,000 vultures at a single New Delhi dump; another shows 200 vultures surrounding one animal carcass, which they could strip to the bone in twenty minutes; a third captures a skinner removing the hide from one side of a carcass while a vulture feasts on the other. Airplane pilots regularly feared collision with these massive animals in the crowded skies. White-backed vultures brought down fourteen fighter jets in four years in the early 1990s.[4] 'They were as common as crows—no one even bothered to count them,' one conservationist said.[5] White-backed vultures were once the most common large bird of prey in the world, their range concentrated in India, drifting across the subcontinent and spilling into parts of South-East Asia.[6] Maybe there were twenty million, or fifty million, or a hundred million. Who would have thought to try to count such plentiful creatures? Why not count the leaves on a banyan tree or the stars in the night sky or the innumerable gods and goddesses in the Hindu pantheon?

By the late 1990s they were nearly extinct.

The animal keeper's question lingered with Asad Rahmani, so the next time he went to Patna Lake in Etah, Uttar Pradesh, where he'd often take his students to study birds, he planned a diversion to the nearby Sikandrarao carcass dump. Looking up into the skies over Uttar Pradesh, he counted the birds, once considered too universal to count, and saw indeed that there seemed to be fewer than he remembered. Later he went to the carcass dump in Meerut to talk to the villagers, and all reported that the vulture numbers were declining.

'Sahib,' one exclaimed, 'Americans have taken them away!'

In New Delhi, one of Asad's former students had become head of the Wetlands Division of World Wildlife Fund, India, (WWF-India), and he too reported that the vultures were no longer breeding on the neem trees that lined the long city avenues, and fewer were seen atop the domed tombs in Lodhi Garden.

The next spring, Asad became the director of BNHS, India's largest and oldest wildlife conservation organization, formed in 1883 in Mumbai (then Bombay) when a group of Raj-era hunters wanted to create a collection of their trophy hunts. Concerned by the reports of diminishing numbers, Asad turned for answers to Vibhu Prakash, BNHS's principal scientist. Vibhu had studied bird populations in Keoladeo National Park, one of India's most famous bird reserves, located in Bharatpur,

Rajasthan, for his PhD thesis. Though Vibhu's main interest was the more charismatic raptors—black-winged kites, sharp-toed eagles and honey buzzards—he had diligently documented vultures as well when he did his fieldwork in 1984. Then, there were 353 nesting pairs of vultures in Keoladeo. When Vibhu returned in 1997, after Asad had warned him of the decline, there were fewer than half of that.

'I saw a lot of empty nests, and when I started looking, there were dead birds everywhere—under the bushes and hanging from the trees, dead in the nests,' Vibhu told me when I first spoke to him on the phone in 2008.[7] 'I was quite worried,' he said, in a grand understatement. 'There were so many it was hard to count them individually,' Vibhu said of the earlier days. 'We'd see hundreds flying and count them by the tens or in groups of fifty.' By 1999, however, not one breeding pair remained at his survey site. BNHS acted immediately, convening a meeting in Mumbai, where biologists from all over the country gathered. They confirmed that the three dominant species of South Asian vultures—white-backed (*Gyps bengalensis*), slender-billed (*Gyps tenuerostris*) and long-billed (*Gyps indicus*)—were rapidly dying off across the region. By 2000, populations were down by an estimated 97–99 per cent, prompting the International Union for the Conservation of Nature (IUCN) to classify all three *Gyps* species as critically endangered, the highest-risk category.

It was extinction in the making.

The Indian scientific community collaborated with international colleagues through the newly formed Asian Vulture Crisis Project to identify the cause of the crash. Scientists initially speculated that an infectious disease or bioaccumulation of pesticides could be the culprit, like DDT had been for predatory birds in Europe and North America. Rumours blamed

Americans—'so technologically advanced', the Indians like to quip—for producing some new chemical that was killing the vultures, or for gathering them up in planes to take to some undisclosed location.

Strangely, even though millions of birds were missing, scientists found it difficult to get their hands on carcasses fresh enough to test. The Indian government had clamped down on biopiracy—the use of natural materials for commercial profit—after a US company attempted to patent a pesticidal extract from the Indian neem tree in the 1980s. These new restrictions meant that, even when biologists had fresh vulture remains, they couldn't export them for testing abroad. Conducting tests within India was unsuitable because of insufficient equipment and accepted standards. Pakistan proved more welcoming.

It would take minds from many nations to unravel the riddle of the vulture decline. Vibhu worked closely with Andrew Cunningham, a veterinary pathologist with the Institute of Zoology in London, but ultimately it was American Lindsay Oaks, a soft-spoken microbiologist at Washington State University who collaborated with the US-based non-profit organization, The Peregrine Fund, to isolate the cause. Lindsay began his investigation in Lahore in November 2000, when he met Munir Virani, a Peregrine Fund raptor biologist from Kenya (and a newly-wed who had to leave his bride at home in Nairobi most of that crisis-filled year), and Patrick Benson, a biologist and American expat living in South Africa.

'As soon as I walked out of the hotel in Lahore, we saw vultures roosting in Lawrence Park right in the middle of town,' Lindsay recalled later from his home in Washington state. 'They were all along the roads outside of town and at the three colonies we went to, there were a thousand birds and nests all over the place. It was the same as seeing a starling here.'[8] Such vulture sightings had all but vanished from India, where the crisis was more advanced. The deaths seemed to be moving in a wave across South Asia, originating in India.

These three scientists travelled to the Changa Manga forest, named after bandits who roamed the area in the British era. Still abundant with vultures, it was bordered by vast agricultural lands in the fertile Punjab sliced in half during Partition. The forest was close enough to India, so that the vultures, which cover vast territory in their aerial meanderings, could have easily flown over the fields where Umendra Dutt's farmers were experimenting with organics. Lindsay wrote later of the wonder he experienced as he watched 'entire vulture colonies, which can contain hundreds of birds, all rising up together and out of sight in a giant column'.[9] At Changa Manga and several other sites in Pakistan, scientists were able to start collecting carcasses of dead birds for testing.

'We were testing everything,' Munir told me when I first met him in person in late August 2006 as he was passing through New York City on business. 'Lindsay was in tears, telling me, "I can tell you why they're *not* dying."' Was it pesticides? Organophosphates were in widespread use for cotton and wheat crops across the Indus Plain. Could it be microplasma, a tiny bacteria? Maybe the hotter temperatures caused by global warming were allowing some bacteria previously kept under control to proliferate. They were trying to think of anything that fit the parameters. It had to be something new. Something universal. Something on the move.

There were repeated trips to Pakistan. In May 2002, the ongoing conflict between India and Pakistan had flared up again when Munir arrived. 'A million troops on the border,' he told me, 'and off I go to collect birds.'

Munir and Lindsay gathered more than 2,500 birds over the course of three years. Most had visceral gout, an untreatable kidney failure that causes a crystallized bloom across the vultures' internal organs as uric acid builds up and coats their insides with chalk. 'We asked what else can cause acute death—heavy metals, pesticides and viruses—and nothing came up in the tests. Maybe, just maybe,' Lindsay said, 'it's some medication going through livestock.'

Martin Gilbert of The Peregrine Fund had set up shop in Multan, Pakistan, and was working with the Ornithological Society of Pakistan and students at Bahauddin Zakariya University. The students surveyed local vets and pharmacies to find out what was being sold for veterinary use. After amassing a long list, they looked at each product to see what fit. 'It had to be absorbed orally, not injected, and it had to be something recent,' Lindsay recollected. 'One drug popped up. At that point we went back to our necropsy samples and did the testing and every single bird that had died of kidney failure tested positive for one drug: diclofenac.'[10]

'I'll never forget,' said Munir. 'I was sitting at my computer in April 2003 and Lindsay called me. Four out of four had tested positive for diclofenac. We did eight more birds and it was a 100 per cent correlation.'[11] At the BNHS office in Mumbai, Asad Rahmani jumped from his chair and rushed to phone Vibhu. The next month, Lindsay presented his discovery at the Sixth World Conference on Birds of Prey and Owls in Budapest. He published the results in *Nature* the following year. Rhys Green, a field biologist from Cambridge University, later found that just one contaminated carcass out of 760 would be enough to account for the population crash they were witnessing.[12]

Diclofenac. It is a non-steroidal anti-inflammatory drug (NSAID) of the sort that humans have used for decades to alleviate pain. Akin to ibuprofen (Advil, Motrin) and naproxen (Aleve), this fast-acting and prevalent painkiller brings comfort to humans suffering from arthritis, cancer or sore muscles. I met raptor biologists who took it for plantar warts pain. Ironically, it also helps alleviate the pain from gout in humans, while causing fatal gout in the vultures. Diclofenac had been restricted as the intellectual property of pharmaceutical titan Novartis, but when the patent expired around 1990, India's generic drug industry, coupled with a thriving black market, flooded the country with cheap, highly-potent diclofenac. After taking it themselves for years, Indians began using it in the early 1990s to ease the aches of their farm animals' cracked hooves and swollen udders. Farmers loved it. Their livestock loved it. Millions of doses were sold. A vulture that fed on the carcass of an animal that had received a dose of diclofenac in the days prior to its death would be dead within weeks.

'Why does it only affect *Gyps* vultures?' I asked Lindsay over the phone, 'and why does it cause gout?'

Lindsay laughed. 'I wish I knew! *Gyps* are just exquisitely susceptible to diclofenac.'[13] Even a residual level—a tenth of the normal therapeutic dose for animals—is lethal for them. 'And we really don't know why. It's kind of unusual. I am unaware of any other creature in which that's true.'

The only hope for the survival of the *Gyps* vultures was a complete and blanket ban. 'You can't cut back on diclofenac,' he said, 'you need to remove it completely. They won't develop a tolerance. There's no active process where the vultures can build up a resistance.'

Lindsay recalled the abundance in downtown Lahore back in 2000. 'Virtually all of those colonies are gone now,' he said. 'I almost don't want to go back; it'd be so depressing.' Whether the vultures return or not, Lindsay will never know. In 2011 an aggressive and rare neuroendocrine tumour ended his life. He was fifty-one years old.

In 2006, the Indian government responded by enacting a ban on the sale of diclofenac for veterinary purposes, but it was already too late. The *Gyps* vulture population had plummeted from forty million to fifty thousand in the decade after veterinary diclofenac was introduced.[14] Seven years after the ban, the numbers hovered at ten thousand. Several hundred long-bills still fly over the cliffs of Ranthambore in Rajasthan. Some perch high on the domed pavilions of Orchha's cenotaphs in Madhya Pradesh. I have seen a colony of twenty white-backs on stick nests in the crooks of trees along a hidden riverbank in Bandhavgarh, but some scientists have started calling these species 'functionally extinct'. That first time I met Munir Virani in New York City, he said they were 'monitoring to extinction'. Munir and his colleagues were biologists, people drawn to the profession whose root, 'bio', means 'life', yet they found themselves surrogating as hospice workers, sitting by the bedside of a disappearing world. All they could do was study the birds that remained, glean what last information they could, and then watch helplessly as the creatures blinked into oblivion.

The vulture collapse was the most catastrophic avian population decline since buckshot wiped out the passenger pigeons that once blanketed American skies, and that slaughter had taken the better part of a century. DDT extirpated peregrine falcons and bald eagles in North America, but similar subspecies continued to thrive in other parts of the world as they recovered. The three species of *Gyps* vultures, in contrast, are limited to the South Asian subcontinent and parts of South-East Asia, and it had taken just a decade for them to nearly vanish.

The disappearance of millions of vultures may not seem to matter to those who see them as harbingers of death, or to those who don't think much about birds at all. But it is an environmental emergency. Nature abhors a vacuum. What will fill this newly vacant niche in our ecosystem?

'Birds are indicators,' Vibhu told me once.[15] 'If something can kill birds, we're not far behind.'

Some believe that we have entered the time of the Sixth Great Extinction, an epoch that the Dutch chemist Paul Crutzen dubbed the Anthropocene. Humans have taken dominion over the planet, transformed its land use and the flow of its rivers, altered the chemistry of the upper atmosphere and the deepest oceans. Many species have not adapted well to our experiment. Of the more than 71,000 species that the IUCN has assessed, 21,286 are threatened with extinction.[16] In India alone, eighty species of birds, ninety mammals, and hundreds of plant species could be lost.[17] It is possible that the three species of *Gyps* vultures that once dominated South Asia could be entirely gone long before the end of the twenty-first century. Will they be missed?

MANU'S ARK

Vibhu Prakash leads a team that spends days and nights coaxing new life from 123 captive vultures in the north Indian state of Haryana.[18] I travelled five hours north of New Delhi by train from the flats of the Indo-Gangetic Plain towards the Himalayan foothills, the haze easing with the elevation gain. At Pinjore, I took a jeep across a wide, dry river bed where fires burned outside of makeshift settlements and continued through a few villages before entering the Bir Shikargarh Wildlife Sanctuary.

The plantation, thick with eucalyptus trees, was once a maharaja's hunting grounds. It now served as a habitat for wild leopards, mongoose, fox and chital deer. Red-rumped rhesus macaques formed a phalanx along the road, waiting to be fed by Hindus worshipping the local incarnation of their monkey god, Hanuman. The sanctuary was also a haven for the Pinjore Vulture Conservation Breeding Centre, where most of the birds that are the seminal stock for the BNHS captive breeding programme live.

On a 5-acre plot leased to them by the government, Vibhu and his wife Nikita, along with a staff of ten, keep an ark of vultures. In the Hindu account of the Great Flood, heroic king Manu, son of the Sun, was told to take the beasts of the world on to a ship for safe keeping while the world drowned. Manu, like the Western world's Noah, became the keeper of a genetic repository. On their small plot, Vibhu and his team make offerings of goats—freshly slaughtered, certifiably diclofenac-free—twice each week to keep their vultures alive into an uncertain future.

In 2001, the Pinjore centre was the first vulture-breeding programme to open. How it ended up in Haryana was luck by chance. Vibhu and Nikita first sought out a site near Keoladeo in Rajasthan, but they found little government support there. Years earlier, R.D. Jakati, head of the Haryana Forest Department and a man with a sweet spot for birds, had called on Vibhu for help. There was a cinereous vulture, a bird of prey with a 10-foot wingspan, kept in a cage as tourist attraction, and he wanted it released. Years later, when he heard Vibhu was looking for a site to breed vultures, he offered the plot in the Haryana forest where they'd released the cinereous.

The breeding programme's initial goals were lofty: to establish six centres, each with sixty of the three *Gyps* species. No one had bred vultures before, and no one was certain that the plan would work. Finding the money, land and staff to support breeding centres proved difficult. But despite the challenges, in 2005, another breeding centre opened in the Buxa Tiger Reserve in West Bengal. A couple of years later a third opened at Rani Forest Range in Assam. The original hope was to begin releasing vultures back into the wild in 2010, but by late 2013, three centres kept only 238 birds, a fraction of the 1,000 birds initially hoped for. News articles bubbled up periodically with cheery announcements of imminent release, but the headlines rarely matched the reality that the centres faced. To release vultures when diclofenac was still present was to send a bird to its death.

I first arrived at the centre in Pinjore in the fall of 2009, just in time to watch the vultures get fed. A research biologist backed his jeep up to an acacia tree and opened the rear hatch as his helpers spread a bright yellow tarp on the ground and hung a scale from the tree branch above. They lashed plastic aprons around their torsos and hooked blue sanitary masks over their ears, but their hands and feet bare but for flip-flops. One by one, the men slid goat carcasses—skinned and slippery, fresh and odourless—out of the vehicle and on to the tarp. Some of the tails, slick with moisture, still had coarse black hair on them. One man stuffed strings of loose intestines back into the animals' abdominal cavities. Another marked the weight of each animal in a notebook as they divvied up the goats between three large aviaries.

Clean meat is the single biggest expense for the centres; Pinjore alone spends Rs 78 lakh a year on goat meat.[19] The vultures eat forty-five goats each week, fifty-five during the long breeding season. Though huge amounts of financing have come from the Royal Society for the Protection of Birds (RSPB) and other Western funders—India more often provides in-kind support through forest lands and infrastructure—lack of funding remains a persistent challenge and threatens the viability of the project. Vultures are pretty much the opposite of what are known as charismatic megafauna: tigers, elephants, whales.

In Pinjore, a pair of assistants wedged two goat carcasses into a single tall blue bucket and then disappeared towards the aviaries obscured behind the trees of the wooded property. In order not to disturb the vultures, they opened a small hatch to slip in the bodies. It was breeding season, and even if some of the birds were accustomed to small intrusions, others might get jittery, and toss eggs or abandon incubation. It is a risk the centre can't afford to take. Female vultures will lay only one egg per year, if they lay at all.

The aviaries were strictly off-limits to visitors, so I headed into an office with Vibhu's wife Nikita, a small woman with a youthful air, to spy on the birds on closed circuit television. With a flick of her finger, she guided and focused the camera. We could scan most of the 100 ft x 40 ft x 20 ft building, constructed around existing trees and open to the sky but for protective netting. Here, she told me, the birds could do everything that a wild vulture could—except, she acknowledged, soar for hundreds of miles as they do when free. She zoomed the camera in on the two fresh goats that lay over a large pile of bones. Gradually, the vultures descended from their platformed perches high up along the walls where they had meticulously built nests out of leaves and sticks.

Ten minutes after the goats had been slipped into the aviary hatch, a lone vulture stepped up to the carcasses. A minute later, there were ten, and a minute after that, the goat was entirely concealed by vultures, wings flaring, heads moving like pistons, feathers flashing in the sunlight. It was an organized yet primal explosion of excarnation. Nikita, a vegetarian, watched entranced. She told me they are aggressive towards the meat but not to each other.

Just ten minutes later, the frenzy subsided. Bones jutted out from half-eaten flesh, and a one-winged bird climbed to the top of the carcass. It had lost its wing at the Ahmedabad Kite Festival, where flyers glaze their strings with glass shards to cut competitors' lines. The razor lines collide with three thousands birds every year, and dozens of injured vultures have ended up at Pinjore.

Most of the satiated vultures stepped away, with hefty grapefruit-sized lumps protruding from their crops, at the base of their throats. One bird approached a water bath nearby, dipping its head down and lifting it up, letting gravity carry the liquid down its throat. And then it sat placidly, broad blunt talons spread wide and its wings and tail falling in an even line just above the edge of the water trough, like a ladies' skirt. The early morning sun glistened off the triangular section of white feathers on its back—for which the white-backed vulture is named—in striking contrast to the dark slate grey of its wings. Its head was tinged with pink. A large silvery beak and dark eyes were set in a brooding brow. Despite their reptilian heads and their leathery skin, they were elegant. A fluffy ring of down feathers encircled its long bare neck.

'So white. Snow-white,' Nikita said softly, smiling. The birds had identification bands around their legs, but no names that she would admit to. The birds remained genderless unless they were caught in the act, since male and female vultures are indistinguishable.

Nikita had studied zoology and put in three years at law school. She had travelled solo to America when she was twenty-one, at her father's behest. He had urged all three of his daughters to explore the world they lived in before they got married, and Nikita had chosen to see New York and the Grand Canyon. On her first day as an administrative assistant at BNHS, she stopped by the law school and learned she had passed her law exam, but she said that she never thought of quitting BNHS.

'It was my destiny to be with BNHS,' she said, smiling. It was there she met Vibhu, and they married a couple of years later. They both did fieldwork in Keoladeo; she focused on king vultures while Vibhu studied all raptors. Most days, Vibhu goes straight to his office and doesn't emerge for hours, while the rest of the staff operate on something closer to Indian Standard Time, taking multiple breaks for tea, chatting and birdwatching. Vibhu doesn't eat lunch or snack while at work. Instead, he takes sips of black tea from a glass and works without pause.

Nikita filled hours of her day observing the birds on the cameras, filling logbooks of their activities. Do you ever get tired of watching them? I asked her.

'Never,' she said. 'They're always up to something, always curious.'

She said they are creatures of habit: they eat, bathe and sun, often in sync with one another.

We watched a white-backed vulture hop to the top of a

perch pole. It stretched out its glorious wings, only 7 feet across—white-backed vultures are the smallest of the three *Gyps* species—and cast a great shadow upon the ground. Before my eyes, the bird became a living totem pole, an echo of the winged Farohar symbol of the religious group the Zoroastrians (known as Parsis in India), their representation of a guardian angel.

I found Nikita's love for the birds contagious, tapping into some ancient connection humans have had to these scavenging birds. Vultures have taken the form of the divine in India and Egypt, their proximity to the heavens making them shaman-like intermediaries between life and death. The Parsis depend on vultures for their last ritual on earth. In the Ramayana, the vulture god Jatayu tries to stop the demon king Ravana from kidnapping Sita, an avatar of the goddess Lakshmi and wife of Lord Rama. Jatayu deflects Ravana's arrows but is felled when the demon slices off Jatayu's wing with his sword, stealing away with Sita.

Vultures' naked-headed repulsiveness is irresistible material for myths about beauty lost. One Amazonian tribe believes that the vulture once had a headdress of elaborate jewels that were a hoarded gift from the gods. A Brazilian hero gouged them from the vulture's head, leaving it bare and returning the glimmering stones to their rightful place in the sky—as the sun, moon and stars. A less romantic Native American story tells of an overeager vulture that finds itself with its head stuck up a buffalo's ass, losing its feathered headdress as it emerges. When the World Cup went to Africa in 2010, vultures were killed, so that spiritual mediums could smoke their brains in hopes of gaining prophetic power to predict the outcome of the games.

To breed vultures in captivity is to enter a new era in the relationship between these mythic birds and ourselves, and it is a tentative experiment at best. Basic biology works against the scientists. Of the thirty-two vultures I watched in the white-backed aviary, there were only twelve established pairs. They build nests for six weeks before the mother lays a single egg. Together, the parents incubate the egg for two months. If the egg hatches successfully, the nestling bird will spend another four months in the nest before fledging with its first flight.

When I visited Pinjore in 2009, only nine eggs had been laid, and only two of those birds had made it to their first birthday. Those that survive need another five years before they reach sexual maturity themselves. The process is slow, requiring patience and dedication. When the same breeding efforts were done to bring the peregrine falcon back from the brink of disappearance in North America in the 1970s, the harmful culprit of the bird's extirpation, DDT, had been banned. And as soon as a young falcon could fly, biologists could set it free. Though now much more closely regulated, diclofenac use is still common in India. Each vulture born in BNHS sanctuaries would have to remain there for years before it could be released.

As I watched these huge, elegant birds—the last of their kind—with Nikita, I experienced something I can only define as a prenostalgia, a tender longing for something that was still right there. I watched them in movement, and the images on the screen before me were live, but there was the possibility that one day, all that would be left of India's vultures was a recorded image. A ghostly register of something that once was.

Outside BNHS's office, a haze clouded the sky. The feeding had stirred the mostly silent vultures. The aviaries were alive with their screeching. They sounded like a hundred creaking doors swinging open and closed—a haunting sound, dim and distant.

Their scratchy racket receded into the background as Vibhu

emerged from his office to lead me down the path along the property's fence line, past the aviaries and beyond an overgrown area where new aviaries would be built to an enclosure hidden in the trees, where a bird called C99 was kept.

A family that lived in a village 50 miles away had found the sick vulture lying in a field and took it in. The family tried to care for the bird, but couldn't understand why it wasn't eating their flour chapatis and milk, so they called BNHS for help.

'It was so weak that you could just walk up to it and pick it up,' Vibhu whispered to me as the long-billed vulture watched us. But once they put meat in front of it, red and raw, it ate on its own and soon regained its strength. It was clear that the bird hadn't been poisoned by diclofenac, or it would already be dead. I peeked at the vulture through the bamboo slats that were woven through the fencing. Slightly larger than the white-backed vultures I'd been watching with Nikita, C99 sat high on a perch that was encircled with coir rope, its belly balanced on the branch, wings relaxed at its sides and bare grey head up and alert.

'It is still more used to humans,' Vibhu murmured, delighted. 'The others would be throwing up in response to our presence right now.'

We turned and walked back towards the office in silence. In spite of the poor salary and quixotic endeavour he was overseeing, Vibhu bore little resemblance to the charismatic alpha males who typically run non-profit organizations and other such enterprises. He was quiet and shy—likely drawn to his work because of its solitary nature. Vibhu grew up in Meerut, Uttar Pradesh, where villagers first noticed the decline in the number of vultures. As a young man, he had been interested in canids—dogs, foxes, jackals. He had just finished with his undergraduate work in 1980, when he wrote to Dr Sálim Ali at BNHS.

Sálim Ali is a legendary name in the world of birds, a man

who fell in love with ornithology when he was a boy of ten in 1906, after he pointed his nickel-plated Daisy gun at what he thought was a sparrow and shot it from the sky. A blush of yellow on the dead's bird's throat ('like a curry-stain', he wrote later) intrigued him and he took it to BNHS to help him identify it.[20]

BNHS's eager assistance had moved the young Sálim and he spent the rest of his life pursuing ornithology, joining BNHS in 1947.

'He wrote back in a week and told me to come for an interview for avifauna,' Vibhu said. 'They were hiring biologists.'

'Dr Sálim was such a warm person,' he continued. 'So enthusiastic. He would be listening so closely to what you're telling him.'

Under Sálim's guidance, Vibhu left dogs behind and turned his attention to the skies. With BNHS, he ringed birds—a process of trapping wild birds and then affixing small identification rings around their legs, to study their movement—near Trichy (Tiruchirappalli) in Tamil Nadu and then at Keoladeo.

Raptors—golden eagles, common buzzards, crested serpent eagles, spotted owlets, black kites, shikras and so many others—captured Vibhu's attention and held it. I recognized his obsession with raptors—I had seen it in others around the world (and was perhaps not invulnerable to it myself). There are birdwatchers, and then there are the raptor-philes.

Vibhu's voice grew urgent as we neared the offices again, rising over the crunch of dried leaves beneath our feet. He spoke of vultures, birds that had never been the primary focus of his work until they started to disappear.

'Without tigers, the ecology can still work,' he said. 'But without vultures, it is very difficult. Nobody can take over their role. They are very efficient scavengers. Nothing will ever be able to fill that niche.'

We had almost reached the office when Vibhu paused by a banyan tree, pointing. It was as wide as it was tall. Its canopy of broad leaves leaned over some of the office buildings. 'Earlier this banyan tree had six or seven vulture nests in it. Wherever you looked in those days, there'd be nests.' He resumed walking, his voice soft again. 'It's so hard to imagine now.'

Some have criticized Vibhu Prakash's work with BNHS, claiming that keeping vultures in captivity is wrong, that BNHS is serving some unidentified self-interest. They argue instead for 'vulture restaurants', where diclofenac-free meat is dumped in open fields to attract wild birds. Vibhu has no patience for such criticism and believes that a restaurant model is ineffective when vultures can easily travel over 77 square miles across their home range,[21] and just one tainted carcass could kill dozens of birds. But the restaurants do get citizens involved. Some have become tourist destinations that attract the wider public, and in that, there is a positive effect. Vibhu knows captive breeding is slow and may be ineffective, but he doesn't see a better alternative.

'We wish we didn't have to do this. It is a big headache to raise these birds,' he said.

Even if the breeding is successful, if BNHS can raise the funds and find biologists willing to do the under-appreciated and under-compensated work of breeding, if the vultures flourish in their new confines, what then? There is little hope for the vultures unless diclofenac is completely removed from the environment. Each year, there will be more vultures to care for. Vibhu, meanwhile, would remain Manu, son of the Sun, a shepherd who waits like Noah for the olive branch to arrive.

A VULTURE IN HAND

'It's a vulture!'

I'm not sure who said it first, or if we all said it simultaneously. It was late March of 2009, and I had just joined Munir Virani and Patrick Benson in Sawai Madhopur, Rajasthan.[22] They were two of the biologists who had helped Lindsay Oaks in Pakistan during the race to find the reason for the vulture decline. Nine years later, they were surveying vulture nesting sites, many of which were inside Ranthambore National Park, where we'd spent the morning peering up at cliffs counting vultures. After lunch, our driver Himmat dropped the three of us off at the end of a road that leads to the Amreshwar temple, not far from town. The plan was to survey on foot, hiking along the ridge outside of the park to look for vultures from above instead of below, and end up at Jhoomar Baori, an old maharaja's hunting palace high atop a hill.

For years, Munir and Patrick have come to survey these cliffs. Munir oversees the South Asian Vulture Crisis work of the US-based Peregrine Fund, and had begun officially conducting surveys of vulture colonies—white-backed and long-billed—in Rajasthan and Madhya Pradesh in December 2002. He would come with a rotating set of assistants and fellow biologists, an 80-pound pack of photography and telescope equipment, and a barely intact roller bag overflowing with spiral-bound notebooks. Within each binder were eight-by-ten photographs, slipped into plastic sheaths, dotted with the thick black markings of a marker pen, of each and every cliff and tree where nests had been sited since 2002.

It was bone dry in this part of Rajasthan in north-central India, a five-hour train ride south of New Delhi. We were between the ancient Aravalli and Vindhya hill ranges, where time had created sheer cliffs, perfect for vulture nests. In a few more months, the monsoon rains would come and revive a deep green to the washed out hues of the dhok-dominated forest, but then it was the rabi season of wheat and mustard. We caught sight of a couple of long-billed vultures perched in their nests as we moved along a path of desiccated leaves towards the Shiva temple, a concrete block buttressed against the base of a steep cliff. It was the second day of spring Navratri, nine days in honour of the goddess Durga, and a man's voice radiated from the temple's tinny speakers, reciting Sanskrit prayers with a hypnotic hint of melody in his voice.

This was tiger country, and there had been recent sightings of T-24, the large male tiger whose territory we had entered. We were accompanied by no rangers. Our only weapons were Munir's 20–60x telescope on a tripod and a fat binder of photos.

'This is the temple of the stone-throwing priest,' Patrick said, as we left the road and walked up the uneven stone steps towards the temple. 'The rhesus macaques can get pretty aggressive here, and each time we come up to this site, the priest is chucking stones at the—'

As though on cue, a troop of twenty or so macaques suddenly started making a ruckus in the ravine down to our left. We turned and saw a vulture standing at the edge of a small pool fed by a rivulet.

That was when one of us, or all of us, yelled, 'It's a vulture!' and also precisely when one of the largest monkeys pounced on the bird, knocking it into the water. Without a moment's hesitation, the two men beside me scooped up stones in their hands and ran towards the bird. I grabbed a stone and scrambled after them.

'Look out for shit,' Virani warned me over his shoulder as I trailed behind them. The footing on the large rocks was unsteady, and the use of the ravine as a human toilet inspired extra caution in our steps as we rushed to the vulture, which was lying immobile in the small pool after the attack. Patrick stepped down, reached into the water, and hoisted up the deadweight of the bird. I threw stones towards the largest macaque, a fierce-looking male that scowled at us, aiming near but not directly at him.

Patrick rested the long-billed vulture down on the ground, its tawny feathers wet across its goose-sized body, blending with the leaves plastered to the grey rocks below. Its bare neck hung limp, but its dark brown eyes were wide open, blinking. It looked at us, alert. Its body seemed paralysed, blunt cream-coloured talons emerged from tough foot pads the colour of charcoal frozen in a tense, outstretched position. Chalky faeces were caked on its back and head, which Munir said is not uncommon with cliff-dwelling birds, whose nests are often situated one on top of the other on a vertical rock face. The two scientists stared at the bird for a moment, assessing the slimy chartreuse mucus issuing from its massive grey beak, and then they got on their cellphones.

Munir looked up the director of the Forest Reserve's number and Patrick dialed, but there was no answer. He left a message. They couldn't leave the bird, for fear of the monkeys, which pulsed around our periphery until we drove them back with more stones. They seemed intent on causing harm to either the bird or us or perhaps both. Next, we called Balendu Singh, owner of the nearby Dev Vilas hotel where we were staying and a long-time supporter of the wildlife that makes Ranthambore National Park such a tourist destination. He offered, generously and immediately, that we bring the bird back to the luxurious hotel. With few other options, we called Himmat, our driver. While we waited, I knelt down beside the bird and lifted it to feel its heft. Though the smallest of the *Gyps* vultures, it still felt close to 12 pounds. I set it back down, my fingers still curled around its neck, which was smoother than I expected, in spite of the smattering of small pin feathers that graced it. It felt thick and soft, yet warm with life. The vulture seemed unable to control its muscles. I rested the vulture's head gently on the ground, at a comfortable angle, and it didn't—couldn't, it seemed—resist. It watched me intently.

The macaques had lost interest by the time Himmat arrived with a cardboard box fifteen minutes later. We drove back to the hotel, where Balendu and an assembly of staff met us. Someone found a larger cardboard box and another lined it with newspaper. Someone else kept the very interested resident hound dogs at bay and another brought an old fruit cocktail tin that Patrick used to pour water down the bird's throat, to keep the sick creature hydrated. He propped open its thick beak and tipped the can, but much of the water dribbled back out, further soaking the damp bird. The staff set up the box in a stark concrete room in an outbuilding, adjacent to a bare washroom with a filthy porcelain sink, and Patrick laid the bird's limp body down on the newspaper, and we walked away, heading back out to continue surveying for any vultures that were still flying by the cliffs at Jhoomar Baori.

It had got too late for a hike, so instead we travelled by jeep. We drove as close as we could to the edge of one cliff, set up our scope, opened our notebooks and set to work.

'Nothing at 4,' Patrick said, his body slightly bent as he peered into the telescope.

'4 had a nest,' Munir responded, pencil poised, looking down at the chart in his notebook.

'Nothing at 4,' repeated Patrick. Munir's pencil made its mark.

'5, empty. 6,' a pause, a squinted eye, '… nothing at 6.'

'6 had a nest.'

'Nothing at 6.'

'7, empty … 7-B high with a chick … Nothing at 8 … Nothing at 9.'

Patrick had been travelling from his home in South Africa to join Munir for these surveys for years. The two interacted like an old married couple. Patrick, at least a decade older than the forty-something Munir, was grizzled and grumpy in contrast to Munir's eternal cheer. The two men bickered, but they were united by a deep love of the animals they studied. They came to Rajasthan in December and again in March, just before and after the breeding season, to count nest sites. Was there a vulture? Was it standing? Building a nest? Sitting, presumably on an egg? Feeding young? They recorded everything diligently.

The work of a biologist is a tedious, painstaking process of close observation. It occurred to me as I watched them that so much of what we know and understand about our world is the result of someone, somewhere, taking the time to stop and observe.

Their work also took them to some of the wildest places in the country—into Ranthambore National Park and up the narrow canyon of the Chambal River and to the cenotaphs of Orchha—and also to some of the most blighted, like the cliffs adjacent to the sprawling cement factories along the Kaimur Range in Madhya Pradesh, where the vulture nests, just a hundred yards away from the road, were obscured by the smoke pouring out of lime kilns fed by loads of stone carried by lines of sinewy women who ascended steep stairs like a human conveyor belt.

I stood behind them and watched. The round ball of a bulbul nest hung from a tree and somewhere nearby, a macaque sounded an alarm call—the sign of a possible leopard or tiger nearby, or maybe just us. My thoughts kept returning to the injured vulture. When I took my turn at the scope, and saw the diminished figures on the cliff through the magnifying glass, I felt that I could recognize the feel of their necks and the way their talons curved ever so slightly, that I knew their smell of damp feathers and fear, tinged with guano. The faraway brought close.

After a few more hours of surveying we arrived at the Castle Jhoomar Baori. We climbed up a narrow stairwell and emerged on the terraced roof of the red-hued structure that was once a maharaja's hunting retreat. The lush foliage of treetops spilled out below us. We could see the last cliffs we needed to survey, blushing red—and vultureless—in the day's fading light. Fairly sure that the sick bird back at the hotel held little hope of survival, we didn't rush back. We sat on the roof and indulged in a cold glass of Kingfisher and a snack of masala papadams as we looked at the empty cliffs. India is becoming a noisier place, but its wilderness is growing quieter.

When we returned to the hotel at dusk, the vulture was still alive. It scurried to its feet when we entered its room, poking its head up over the edge of the tall box, watching our every move. It then promptly fell back down on its feathered rump. Not only was it still alive, it seemed like it was regaining motor control. We began calling the bird JB, after the Jhoomar Baori cliffs where we'd found it, and emailed Lindsay Oaks in the United States and veterinarian Dr Igal Horowitz in Israel for help, listing its symptoms: the frothy, green phlegm at its beak, its inability to fly, and that its body felt stiff.

The next day, the group's consensus was that the bird was probably suffering from organophosphate poisoning. The vulture could have fed on a poisoned animal, baited by farmers who feared local predators, or absorbed toxins dermally through its feet while in fields sprayed with the pesticide. The doctors' advice was to wash its feet and beak well and give it a shot of atropine sulphate as an antidote. By evening, Patrick held JB, who put up a healthy fight, and plunged the needle into its feathered frame.

By the next morning, the vulture was attempting to climb out of its box. I warily eyed the spinning ceiling fan not six feet above the (potentially) flying creature, and then looked around the room. There was a simple cot and a few staff uniforms hung on pegs on one wall. What I had first mistook for a storage room was actually part of the staff quarters. I'd stepped through the thin veneer between served and server, yet I still didn't know who had slept with this vulture through the night, smelling its excretions and listening to its large wings scraping against the sides of the cardboard box.

Patrick and Munir had replaced the tin can with a large syringe to direct water down the vulture's gullet. They used their fingers to stuff raw chicken down as well, massaging the meat down into the bird's crop.

By day three, JB was eating voluntarily. Its neck regained strength. When we brought the bird out on to the lawn, it ran to hide in the shrubbery, though it was still unsteady on its legs and tipping forward under the momentum of its own movement. We had to leave the next day, so a local man agreed to keep JB in an enclosure until it fully recovered. A week later, Munir called to check up on JB's condition.

'It was released three days back, sir,' the young man told us over the phone. 'It was doing well and it flew off.' JB had returned to the wild.

Nine months later I returned to survey the cliffs with Munir and Patrick again.[23] We were up at dawn on the first morning, Himmat driving us into Ranthambore National Park, past the crumbling remains of temples, through an elaborate stone arch, and under banyan trees whose roots cascaded down rock walls like water. Above us in the clouds was the historic Ranthambore Fort from the tenth century, perched high on a hill, the park encircling it like a wild, forested moat.

We were in a forest alive with the cacophony of rose-ringed parakeets and alarm calls of the lanky black-faced langurs that hung from trees. As we bumped along in the jeep heading to one of the dozens of cliffs they surveyed, Munir recalled his first trip to Ranthambore in April 2000, when he witnessed a hundred vultures on just one cliff, yet also a dozen dead and dying.

At the same cliff, we now counted six long-billed vultures' nests where there had once been a hundred, and a few birds were soaring on the first of the day's thermal winds. On page after page of the survey notebook, Munir marked down neat 'e's—one for each empty nest—in the December 2009 column as Patrick reported from the scope. We saw a still figure at one spot on the cliff. It was a dead vulture. In the five years between 2004 and 2009, Munir recorded nineteen dead birds. He also watched as all the white-backed vultures vanished, leaving only some long-billed vultures, not just at Ranthambore, but at survey sites throughout Rajasthan and Madhya Pradesh.

The cliffs we were looking at were part of the ancient Aravalli Range, part of the same mountain chain that encircled Kanhaiya's village of Bhaonta. They rose vertically into the empty sky, exposing striations of geological time. Somewhere in those cliff faces was embedded another era of extinction, from the Cretaceous–Tertiary period. Approximately three quarters of plant and animal life perished around sixty-five million years ago when, it is believed, an asteroid collided with the earth. Now, humans are creating our own destruction.

But there is still life abundant. On our surveys we saw red-headed vultures, also known as the Asian king vultures, black-shouldered kites, and Bonnelli's eagles soaring high. Tiny spotted owlets and scops owls peered from tree holes. There were tigers too. Magnificent beasts that walked through the forest like kings and queens, ignoring the safari jeeps filled with tourists that pursued them with cameras, ooos and ahhs. On one of our last days in Ranthambore, Munir, Patrick and I stopped by a small lake and watched the bountiful bird life around the water. There were bar-tailed godwits, white-throated kingfishers, pintails, green wagtail, cormorants and swooping swallows. Peacocks strutted along the banks. Black-winged stilts kicked up water with their red spindly legs.

We opened a thermos and poured out small cups of steaming tea.

Both men mentioned the breeding work of Vibhu Prakash at Pinjore. Vibhu said that human diclofenac had to be banned completely in order to eliminate the black market for veterinary use.

'I agree with him 100 per cent,' Patrick said. 'Unless you ban human diclofenac, the problem will never go away. There will always be a black market.'

That's what need be done, but is it feasible?

'No, no,' said Munir. Though, by 2014, Munir and his team, along with Vibhu and other researchers, were seeing hints that the vultures' precipitous decline had levelled. At some sites, there had even been a hopeful upward tick, but the numbers were still incredibly low.

'When we started the surveys, we were looking at the dregs,' Patrick said. 'It's like having a thousand birds and going down to ten before you start getting data. And so we're up to twelve. Maybe.'

'Overall the trend is down,' Munir agreed, 'but my feeling is that after the ban, the decline has slowed down.' But it was still dire. White-backed vultures had once far outnumbered the long-billed vultures. 'Now,' Munir admitted, sipping tea and looking out across the water, 'they're completely gone.'

The conversation, in spite of the bucolic setting, had turned dark.

'It goes back to the same thing I've been saying over and over,' Patrick said, refusing to see an upside. 'We're a failed experiment. There are just too many of us and no one is talking about it,' he said, speaking of the impact of more than seven billion humans on the planet, of the way our actions have affected every recess of land and water, right down to how our drugs could wipe out a species.

'We're soiling our nest,' he said.

Silence hung over the jeep.

I asked them if any other birds could fill the niche the vultures had vacated. Munir said the niche was being filled, but it wasn't by birds.

'Go to Bikaner, in western Rajasthan,' Munir told me. 'Go to the carcass dump and you'll see.'

THE DOGS

The vultures are gone, but the livestock carcasses they once consumed by the millions remain. Many are collected and deposited at carcass dumps like the one called Jorbeer on the outskirts of Bikaner, where dogs run wild amid an endless supply of food.

As I travelled around India, I kept hearing about aggressive dogs. Soon after I arrived in Bikaner,[24] someone told me about two local girls, eight or nine years old, who were attacked by dogs at night, while they were sleeping. They were such easy prey.

'They were hurt so badly, but not killed,' the man told me. 'The police came and took the dogs away, but I was so astonished … how can there be dogs like this?'

Other dog-attack victims—seventeen million in India each year, mostly children—do not survive.[25] A four-year-old boy named Manjunath in the city of Bengaluru didn't make it. Nor did three-year-old Arshpreet Kaur, who was bitten in her home in New Delhi when a stray dog came through her front door; it bit both her and her grandfather.[26] First Arshpreet got a headache, then a fever. She slipped into a coma where she remained for nine long years before she finally died. A journalist reported that dogs routinely enter the Patna Medical College in Bihar, where one had to be threatened extensively with a baton when it was found with the skull of a stillborn baby locked in its jaws in the gynaecology department.[27]

Munir had warned me of the dogs, and they were there when I arrived at Jorbeer and met Rameshwar and his wife Pandevi.

'I've been here for four years,' Rameshwar told me as we stood outside his family hut, a hundred yards from a 5-acre pile of carcasses. He was wearing a thin button-down shirt that had a neat tear in the back. His hair was thick, as dark as his onyx eyes. Lanky, he looked no older than thirty.

'I live here with my wife and four children,' he began. 'These are my four goats, which we keep for milking. Every day from the city, a tractor comes with thirty to thirty-five carcasses …' His gaze lifted up as something captured his attention behind me.

'Heyyahh!' he yelled, his soft voice suddenly rising, as he shooed goats from the open doorway of his hut. They scurried away, but remained close. I asked him about the dogs.

'I've seen the dogs coming, each day, more and more,' he told me. 'During the day, they are very familiar with us, but at night, they are much more aggressive. Then, we don't go outside without a stick, and I have to keep them away from the goats.' He motioned to an enclosure made of thorny brush, where he put the goats at night.

The sun blazed on us and the dusty desert ground, littered with a goat hoof, a stray tail, a single shoe, and the plastic bags that have become an integral part of the Indian landscape. We were upwind from the bodies, so the air didn't smell fetid.

Bikaner is a city of a half-million that sits in the Thar Desert of western Rajasthan. Though it had been dry amid the cliffs of Ranthambore in the eastern part of the state, here were no forests, just expanses of sand dotted with occasional trees. Once, the government of India grazed a huge camel corps at the Jorbeer site, until the military started adopting more modern means of transportation. But dozens of camels, long-legged dun mountains, still grazed on the surrounding lands, blending in with the tawny hues of the sand dunes, upon which grew gum acacias and khejari trees, Sodom's apple and shrubby khimps. Gazelles, chinkara, desert foxes and porcupines still roamed the area. But since the camel corps had dispersed, the carcass dump had appeared in its stead—no one seems to know exactly when or how or by whose directive. It was a convenient place for the city to bring its dead cows, water buffalo, goats and camels, as well as some dogs I saw tossed from the back of a boxy yellow truck before it turned and sped away, kicking up a cloud of dust.

Rameshwar lived at the dump as the on-site skinner, a profession mostly left to the Dalits. When the carcasses arrived twice each day, Rameshwar and his wife Pandevi removed the hides, to be sold for leather, and then piled up the bare remains for the animals and elements to dismantle. They layered the skins with a desiccant in covered sheds at the side of their encampment, which consisted of a thatched hut that tilted to one side and an outdoor earthen platform with a traditional clay chulha for a kitchen. Their eldest son, at thirteen, worked as a labourer on the tractors that brought the deliveries. Their daughter lived with them, and their two younger sons were in and out of school, sometimes staying with the grandparents in town.

Over our heads, hundreds of birds kettled in slow circles in the sky. Of the sixteen Old World vulture species, nine are found in India. Eight of those can be spotted in the Thar. But overhead were other birds—mostly Eurasian griffons, bulky steppe eagles, (which could have travelled from as far as Siberia)[28] and Egyptian vultures the size of large gulls. They all rode the warm whorl of desert thermals to the apex of the gyre without a single flap of their wide wings and then peeled off like a slowly cascading waterfall. The white-backed and long-billed *Gyps* vultures, hardest hit by diclofenac, were noticeably absent. In the prior six years, only one had been seen at Jorbeer.

On the ground, more birds adorned the few scattered trees that drew life from the desert, and others vied with the dogs for the fresh meat of the newest arrivals to the dump. The carcass dump had become a prime birdwatching spot, our places of refuse fast becoming the wildlife refuges of the future. There were fork-tailed kites, cawing crows and a few cinereous vultures. One landed in a treetop, scattering the eagles and griffons on the branches below. I saw the slender white wisps of cattle egrets—a bird more commonly seen riding the back of a water buffalo in a rice paddy—standing inside the remains of a massive bovine rib cage, picking at the leftover flesh. There were drongos with long forked tails, and hoopoes with black-tipped fanned crests like some Indian punk mohawk and cooing mourning doves and clusters of black ibis—one of which had its long, downward-sloping bill buried in an unidentifiable body part of a carcass. Skeletons of livestock past were piled 15 feet high in places. Some but not all of the meat was picked clean, awaiting the bone collectors' arrival. Two Egyptian vultures could barely wait for a dog to finish defecating before they gobbled up its faeces.

Next to Rameshwar's lean frame, his wife Pandevi was full and round. Her voice was raspy, and she smiled easily. Her dark skin contrasted with the bright orange paisley dress she wore and the orange scarf draped over her head. Matching bangles ringed her wrists. A solid gold flower adorned the side of her nose, and a black circular bindi graced her third eye.

Forest officials who lived far from Jorbeer insisted that there was so much available food provided by the daily delivery of carcasses that the dogs had no reason to leave the dump and be 'naughty and mischievous'. Pandevi, who lived with the dogs, disagreed.

'Many go and roam, 2 or 3 kilometres from Jorbeer,' she said.

'In the late night, I am very afraid of the dogs. If I have to go out at night for the toilet, I take a stick,' she said. 'During the day, we'll carry a stone, but most of them know us and it's usually okay. But in the night, and when they are in the mating season, they are different.'

'If we are off working, and there is no one with our animals, and there have not been fresh carcasses, they will attack the goats,' Rameshwar added. 'A few months ago, the dogs killed two of them.' I looked at the four remaining goats that hovered behind him and the weak fortress of thorny branches that protected them at night.

'I've seen the number of birds go up,' he continued, 'but I've also seen the dogs coming, each day, more and more.'

One forest official told me there were no more than 150 stray dogs at Jorbeer. Rameshwar said there were 2,000. The true figure lies somewhere in between. I tried to make my own count, but in the midday heat, the dogs were scattered, seeking shade under the desert berry bushes and acacia trees, ten here, three there, twenty vying for a spot around the newest carcass, chasing away birds that came too close. A female dog bared her canines at a couple of other worn females, defending three young pups that frolicked among the carnage. Even as I passed, dogs looked up and growled from a hundred feet away. Most of them seemed strangely healthy, thick and muscular, unlike the cowering, scrawny strays I'd grown used to seeing on India's streets. But some, their sagging skin and ribs exposed, had lost their fur to dermatitis, which has been moving through the population, killing up to 40 per cent of the pack according to one local biologist. The fatal disease has begun to show up in the wild gazelles that also pass through Jorbeer. The mortality rate is high among feral dogs—dermatitis accomplishes an extreme form of dog control that the government is unwilling to perform—but the dog population was still increasing. There used to be ten dogs to every hundred vultures. The ratio has now flipped.

'Since 1992,' Vibhu Prakash had told me back at the breeding centre in Haryana, 'BNHS has surveyed for dogs as they do their vulture counts.' He hasn't had time to analyse the data, but his anecdotal impression is incontrovertible. 'At the carcass dumps—Jorbeer, one near Jodhpur, Gazipur near New Delhi, Tonk in Rajasthan and Devnar near Mumbai—at all of them I've seen vultures go down and dogs go up.' In the decade of major vulture decline, from 1992 to 2003, one estimate shows dog populations increasing by a third, up to nearly thirty million.[29] Though there is no certain way to show there is a direct correlation, the escalation of the dog population corresponds perfectly with the disappearance of India's vultures.

'Just the day before yesterday,' Rameshwar told me, 'a vulture was eating a cow carcass, and the dogs came and attacked the vulture and killed it. It doesn't happen often, but they can catch the wings of the vulture and then the vulture will die.' Somehow Rameshwar managed to retrieve the dead bird from the dogs, and place it high atop a shrub, in case the forest managers needed to see it. It was still there on my visit, lying face down in the foliage. It was a Eurasian griffon, whose once-fluffy collar of buff feathers was now flattened with death.

Vultures are necrophages; they have no interest in the living. Like some of my Buddhist friends, they are carnivores but they do not kill. Vultures clean up carcasses in a way that dogs and other scavengers simply cannot. They are hyper-efficient in their jobs. After skinners strip the hide, vultures pick the skeleton clean for the bone collectors who gather the skeletons, which will be turned into fertilizer for crops or gelatin for food. Other scavenging birds and dogs prefer eyeballs and internal organs, leaving much behind and making the task of the bone collectors, already unpleasant work, that much more difficult, the residual meat still rotting in place.

But the most dangerous shift in the ecological landscape is that dogs have a proclivity for targeting the living: livestock, wild animals, small children. A veterinary clinic near Jorbeer was treating several animals each day for dog bites.

Dogs, unlike vultures, are hunters. They will scavenge, but their instincts are predatory, especially when the spirit of a pack possesses them. At Jorbeer, I watched them pursue wild gazelles across the sand. Their bodies lowered into a crouch, they moved in pairs to encircle their prey. Their pace quickened, until the gazelles skittered away.

India accounts for 36 per cent of the world's rabies deaths, mostly affecting the poor—and the numbers are increasing.[30] Every year, more than twenty thousand people die in India due to the disease, most of them children who have had contact with infected dogs and are treated too late.[31] In Mumbai alone, eighty thousand people were bitten in one year. In 2001, a national law codified India's religious aversion to euthanizing dogs, and sterilization programmes remain haphazard.[32]

'India's forty million or so vultures were between them eating about 12 million tonnes of flesh each year. With no vultures to clean up, there was an explosion in the population of wild dogs,' author and environmentalist Tony Juniper wrote in his book, *What Has Nature Ever Done For Us?* 'More dogs led to more dog bites and that caused more rabies infections among people. The disease killed tens of thousands, in the process costing the Indian economy a figure estimated in the excess of $30 billion.'[33]

Juniper estimates that the loss of natural services performed by vultures and other human assistants costs the global economy more than $6 trillion per year, equivalent to around 11 per cent of world gross domestic product (GDP). By contrast, he figures conservation of species threatened by extinction could be accomplished with only $76 billion.[34]

The math regarding causation is more slippery than many biologists find comfortable, but increasingly, scientists, citizens and policymakers are acknowledging the value of these ecosystem services drawn from natural resources, biodiversity and healthy ecosystems. The Economics of Ecosystems and Biodiversity (TEEB), for example, is a global initiative focused on drawing attention to the economic benefits of nature that humans siphon from as though it were infinite. It recognizes, and attempts to quantify, what the intact forest that provides a cancer cure or the wetlands that filter drinking water are actually worth to us, assuming we had to pay for a human-made replacement.

By consuming the dead, vultures help keep such diseases as tuberculosis, brucellosis and foot-and-mouth disease contained by inactivating these pathogens, some of which can remain transmittable for months after a host organism's death. A combination of strong stomach acids and body temperatures over 100°F mean vultures can even ingest an anthrax-infected carcass and suffer no ill effects. For the vultures to be unaffected by some of the most virulent diseases of our time, yet killed by a drug as seemingly innocuous as aspirin, is no small irony. The fear is that with vultures disappearing and human handling of dead livestock on the rise, there will be nothing to prevent diseases from spreading among both animal and human populations.

India is home to regular outbreaks of anthrax.[35] The disease is transmitted by spores about which little is known except that they can lie dormant for decades before infecting human tissue. In 2009 and 2010, fourteen heroin users died in the United Kingdom from anthrax.[36] It's suspected that bone meal derived from an infected animal was used to cut the smack. A few years before that, a drum maker who lived in a quiet English village died after contracting anthrax from the imported goat hides he used for his drums, the beating releasing the spores into the air and his lungs.[37]

'When we try to pick out anything by itself,' wrote the naturalist John Muir, 'we find it hitched to everything else in the Universe.'

A couple of days after my first visit, I returned to Jorbeer with Jitu Solanki, a local biologist who is the type of man who can spot an owlet in the nook of a roadside tree while driving at 40 miles per hour down a narrow desert road. There wasn't much paid work for professional naturalists in India, so he opened a guest house, named it Vinayak and began leading desert safaris to make a living. One of his favourite birdwatching spots was Jorbeer. He told me that he was comfortable around the dogs but he was protective of me as we stepped out of his small car, which he parked on the edge of the carcass pile. We stood together with binoculars as he identified the different birds we were seeing. When the dogs suddenly roused and started barking, gathering together in anticipation of something we couldn't detect, he paused mid-sentence and watched.

'Do you worry about the dogs?' I asked, questioning his earlier dismissal.

He said, 'Yes.' He estimated that there were up to a thousand dogs at the site. 'Dogs are a big problem. They are really too much. A few months ago, they came with tractors and …' His voice trailed off as he made a quick whistle and a universal scissoring gesture with his fingers to indicate neutering. 'They put ID tags on them, so maybe in two years it will work. But they never kill dogs here.' He held his binoculars up to his eyes and scanned the birds above us.

'Hindu people, you know,' he continued, 'there is a lot of god and all. We have a god we call Bhairava, an incarnation of Shiva, and his vehicle is a dog; so people believe that if you kill the dog, Bhairava will be angry. This is a very nice concept. In Hinduism, if you see any god, you will find some related bird or animal and it is a very nice way to conserve wildlife. Even the vulture that people relate with death and don't like, this bird is the Hindu god Jatayu, who tried to save Sita when she was kidnapped. So even the ugly vulture has a place with the gods.'

When I returned to Bikaner, I wanted to shake off the afterimage of my time at the dump. I headed to an open-air market to wander and find some fruit. On my way out of the guest house where I was staying, there was a homeless family stationed along the street, a man and woman with five children. I bought fat red pomegranates, roasted peanuts, still warm, a spicy snack mix, and rasgulla, the sticky syrup leaking out of the plastic bag that contained them. As I returned to the guest house gate to share some of my food with the family there, the oldest daughter, maybe eight years old, caught my eye. She swooped in, grabbed the food from me and quickly returned to her circle, where the other children ringed around her. She held the bags close to her chest, guarding the food, though it was little more than a weak appetizer. The kids needed a thali each, mounds of rice and hot chapatis, a variety of sabzis and thick dal. They needed samosas and chaat and pani poori and Thums Up cola. Instead, they had to survive off of what society discards—our excess. I thought I had left the dump at the edge of town, but its ecosystem was mirrored on the streets of every city. India's poor, like its vultures and dogs, depend on what the rest of us can do without.

THE SKIES ARE EMPTY

Far from the streets of Bikaner, in a posh part of Mumbai, human bodies are handled more delicately than the animal carcasses at Jorbeer. The recently departed are carried by people called khandhias to stone towers built explicitly for the purpose of laying out the dead. Sky burials were once the perfect disposal system. Since the days of their prophet Zarathustra in 600 BCE, Zoroastrians have used dhokmas, or Towers of Silence, to dispose of their departed. In their native Persia, they placed bodies on natural stone promontories exposed to the sun, and vultures would descend to feast in a practice called dhokmanasheeni. When Zoroastrians migrated to India in the eighth century CE, escaping persecution and becoming known as Parsis, they continued the tradition on 57 acres of forest land known as the Doongerwadi, a hill grove in the heart of Malabar Hill, one of Mumbai's most upscale neighbourhoods. The bodies are left exposed on top of the towers, mottled with moss, for wild vultures to consume, a task they could complete in mere hours. The burial custom is so ancient that Herodotus noted it in *The Histories* in 450 BCE.

Seeing the towers was part of a regular tourist circuit in the nineteenth century. American author Mark Twain visited Malabar Hill on a round-the-world trip that he chronicled in his book *Following the Equator*. 'On lofty ground, in the midst of a paradise of tropical foliage and flowers, remote from the world and its turmoil and noise, they stood,' Twain wrote of the towers. 'The vultures were there. They stood close together in a great circle all around the rim of a massive low tower—waiting; stood as motionless as sculptured ornaments, and indeed almost deceived one into the belief that that was what they were.'[38] In the *Hindoo Patriot*,[39] he wrote, 'One marvels to see here a perfect system for the protection of the living from the contagion derivable from the dead.'[40]

The Parsi belief system is rooted in the principle that people should live in harmony with nature. 'In Zoroastrianism, death is not seen as the work of God, it is seen as the temporary triumph of evil,' said Khojeste Mistree, a Zoroastrian scholar and trustee of the Bombay Parsi Punchayet, the religious governing body in Mumbai. 'That means that when a person dies, within a few hours, the body is deemed to be ritually impure. And because it is deemed to be ritually impure, we cannot bury the body because that pollutes the earth. We cannot burn the body because that desecrates fire. We cannot drown the body because that sullies the waters. So from a very ecological point of view, the only method available to us is the exposure method.'

So while the vast majority of humans on earth choose to bury or burn their dead, Parsis allow neither. As Mumbai transformed into a modern cosmopolitan city around the forested enclave atop Malabar Hill, time and seasons of monsoon rains sprouted high-rise and high-end apartments like weeds, but the majority of the Parsi population continue to consign their dead to the towers, exposed to the sun and, now, a vultureless sky. After three thousand years, it took less than a decade for diclofenac to wipe out the vultures, slamming the entire system of sky burial into dysfunction.

Today, the Parsis are themselves struggling to survive. While India's population more than tripled in the last sixty years, the Parsi population fell by more than a third, down to 69,000 in 2001, since the most traditional in their ranks adamantly disallow conversion or intermarriage. Some speculate the number could be as low as 45,000 in 2014. For each Parsi born, three perish.[41] And those who die in Mumbai, where the majority of the world's Parsis live, are likely to opt for a sky burial at the towers.[42]

The first time I met Dhan Baria, a tiny seventy-year-old Parsi woman on a mission, she wore pants and her curly hair was pulled back.[43] We first met at the office of the Nargisbanu Darabasha Baria Zoroastrian Foundation, a charity organization she founded in her mother's name, located in a school building on Princess Street. She had oversized features—big ears, big eyes, and large gnarled hands that she slammed on the table for emphasis. Excitable, sometimes angry and just as often gleeful, Dhan spoke in a booming voice. Her skin was fairer than mine, and she was unadorned but for a single silver watch that slipped along her wrist when she gesticulated.

Born in Surat, Gujarat, Dhan moved to Mumbai when it was still Bombay, as a girl of six or seven. She remembered Kemp's Corner, a section of Malabar Hill, long before the high-rise boom. It was a 'lonely place'. And she remembered the vultures.

'There were many vultures,' she told me as we sat in her office. 'When my sister expired, and when my mother's sister, aunty, expired, at that time there were so many vultures here,' she said. 'That was a long time back. My sister died in 1957. Aunty and others passed away twenty, twenty-five, thirty years back, and at that time there were vultures.'

And then in 2005, her mother expired. She decided to follow Parsi tradition and made a phone call to the Punchayet. The pall-bearers—khandhias and nussesalars—carried her mother away to the towers. She followed religious mandates, and returned to the grounds to pray for her mother's soul for months afterwards.

'When I was at the Doongerwadi, praying, I asked the nussesalars if my mummy was gone, and their response was laughter!' she said. '"No, no," they told me. "You're mother will be there for years to come!"'

Without the vultures her mother's body, along with many others', remained exposed and naked atop the towers, slowly rotting.

'Everyone used to notice [there were no vultures], but they

never thought about what happened to the bodies when the vultures don't come,' Dhan said. 'Arrey, can such a big body be finished by sunrays alone? They all were sleeping! There was talk of the bodies not decomposing, but who wants to think of such things?'

Wanting to know the truth, Dhan hired a photographer to sneak into the towers and document what was happening in 2006.[44] She sent those graphic images around on flyers to Parsis throughout the city, and the photos eventually wound up on CNN. There was an uproar. The Punchayet had assured everyone that even without vultures, the corpses were decomposing. The photographs proved otherwise. Bloated bodies, sprawled about, were missing bits picked off by the crows and kites that still came, but were otherwise largely untouched.

Dhan Baria rose from her office chair, and offered to take me to the grounds of the Tower of Silence.

More than a century after Mark Twain's visit, the Doongerwadi remained as he had described them—a paradise of tropical foliage and flowers. It was an oasis, an escape from the toiling, hazy city that hemmed in from all sides. Dhan led me in—we honoured custom by covering our heads, but disregarded that I shouldn't be there at all. The grounds were consecrated and now reserved for Parsis. We walked up the same path that Parsi families tread as they follow their loved ones for the last time, before the khandhias carry the bodies to the towers. As the path wound uphill, she pointed out one of the structures through the leaves of the lush forest. I could just make out the form of the squat basalt cylindrical structure, 40 feet across but only 23-feet tall as I later learned. It appeared like a mirage within the forest where butterflies flitted about and peacocks sat picturesquely on the low branches of trees.

We stopped as the path ended at three marble slabs where the final prayers are said. It was such a lovely setting, under the wide canopy of forest. The sacred space was dappled with sunlight and bird droppings. A colony of fruit bats dangled from the upper branches of a nearby tree, chittering. There hadn't been a vulture to the site in many years, but black kites called while circling in great numbers overhead. Every day, an average of three more Parsi bodies arrived for them. They came to eat as the vultures had, but their consumption was incomplete. The bodies piled up. Only eyeballs, genitals and small appendages were devoured. There were rumours that fingers had shown up on the balconies of the luxurious residencies that surround the Doongerwadi. People complained of the smell.

Beyond the marble slabs was an elaborate metal gate that stood between stone supports and sealed by a large lock. Dhan, a mischievous look upon her face, directed my attention to a section of the fence that was peeled back just enough for a person or two, perhaps even with photography equipment, to slip through.

She also pointed out all the bare ground that could be used as a cemetery. A faithful Parsi, she lamented the vultures' disappearance, but she believed it was time for Parsis to adapt. She railed against the hypocrisy that Parsis who lived overseas could be buried abroad and have their final rites performed at the traditional fire temples in Mumbai, but local Parsis who chose cremation or burial would be ostracized. '"Mr So-and-so expired,"' she said, of the calls that came to Mumbai from overseas, '"so please start the ceremony there." When someone dies in London, his ceremony can happen here at the Tower of Silence, so if he dies in India—why can't it happen?'

The Punchayet had unsuccessfully attempted to replace the service that the vultures provided with questionable technological fixes. Chemicals had been poured into the orifices of the bodies, but they spilled on to the stone, making it slippery, which was a problem for the khandhias' work. An ozone machine was supposed to help alleviate the smell of rotting corpses. Someone once reported a mysterious fire emerging from the stone tower, suggesting that the Punchayet was burning the bodies in secret. The most effective method thus far seemed to have been the installation of solar reflectors atop steel scaffolding planted adjacent to the towers and directed at the bodies. They speed up the process of decay without violating a fundamental tenet of their religion, which dictates that corpses cannot be burned. Imagine a child with a magnifying glass aimed at an ant. The reflectors, which don't function during Mumbai's four-month-long cloudy monsoon season, desiccate the bodies. The most orthodox priests claim that it is a back door to cremation. Khojeste Mistree, the Zoroastrian scholar, agreed.

'Khojeste,' Dhan told me, 'is a rascal of a person.' She gossiped to me about how he refused to honour his mother's wish for a cremation. His mother had to flee him for the United States, so she could die in the care of her other son. But I found Khojeste to be quite pleasant when I met him on a separate visit to the Doongerwadi.[45]

I stood at the same spot by the gate with Khojeste as I had with Dhan Baria. Khojeste's face was unlined, unmarked by self-doubt. In conversations at the Doongerwadi and at his home in a nearby Parsi colony, he told me about his idea of true Zoroastrianism.

'People say the Towers of Silence are antiquated, that we should move on to cremation and forget our tradition,' he told me, speaking in a precise Cambridge accent though it had been thirty years since he studied there. 'I'm totally opposed. This powerful, vociferous minority of reformists doesn't know the religion.' He clarified that he wasn't a priest.

'I teach the priests,' he emphasized. I found him pleasant for a puritan.

When I first met Khojeste in 2009, he told me about his vision of a vulture aviary where the lost tradition of his people could be regained. It would stand 65 feet high, the size of two football fields arching over the towers and the trees 'without cutting a single branch'. Parsis would not use diclofenac in their dying days, and the caged vultures would be able to feast as before.

Khojeste even went so far as to suggest that some Parsis were keeping vultures, though it is illegal.

'It's a perfectly workable idea as far as I am concerned,' Khojeste told me, 'and it's terribly eco-friendly as well. It religiously fulfils why we use the Tower of Silence, namely: it is egalitarian, it is eco-friendly, it is cost-effective and, most important, from a theological point of view, it is right.' We sat in his office, the blind eyes of his golden dog Sammy shut tight with sleep as he lay on the floor between Khojeste and me. The walls were covered with framed photos and old maps. Etched on a sliding glass door was the Zoroastrian symbol—a creature, part man and part bird with plank-like wings that could be a vulture's. It was the same image I had seen at the breeding centre in Pinjore.

'Really, we are more concerned than anybody that the vultures survive,' he said. 'Our interest is in the very preservation of our religious tradition. We're more invested than those armchair conservationists,' flinging a sideways barb at the biologists of BNHS.

When I returned to see him four years later in 2013, he had partnered with the armchair conservationists of whom he spoke, including Vibhu Prakash and Asad Rahmani.[46] His plans for the aviary, which I had taken for a quixotic fantasy, were moving forward, at least on paper. BNHS, whose conservation goals were unrelated to Parsi traditions, hoped to establish another breeding centre in Dahanu, Maharashtra, for white-backed vultures that could supply the Parsi aviary, but only after a trial year with non-endangered Himalayan griffins eating the bodies.

Critics, and there are many within both the Parsi and scientific communities, say an aviary would lead vultures to their deaths. Diclofenac is present in hundreds if not thousands of human painkiller formulations. It would be virtually impossible to ensure that the Parsis were, as the goat meat at the breeding centre in Pinjore, diclofenac-free.

The Parsi community was divided on what rituals and codes now governed them in the twenty-first century. In late 2013, a criminal complaint of financial impropriety was filed against the Punchayet chairman. All discussion of the aviary fell to the wayside.

'We Parsis love controversy,' said a man named Homi Dhalla when I sat down with him one evening in his home in Colaba, on the southern tip of Mumbai.[47]

'I am neither an orthodox nor a reformist,' he explained, 'but a pragmatist.' Homi Dhalla was the founder and president of the organization called the Cultural Foundation of the Zarathustra World, but also an ardent supporter of interfaith reforms, matrilineal intermarriage, and the freeing of Tibet. He had met the Dalai Lama, the Pope, Mikhail Gorbachev and Nelson Mandela. For thirty-four years, he served on the Doongerwadi Advisory Committee.

'I was just walking on the premises of the Tower of Silence, and I realized that there were no vultures,' he said, sitting in a high-backed chair in his apartment recalling the late 1990s. 'What the hell are we going to do? And then suddenly it was a flash—of intuition, or whatever you call it; I was told that it was solar energy that would solve the problem.' So, by 2000, he had solar reflectors installed on three operating towers at Malabar Hill.

Homi alluded to an even better system in the works, but refused to reveal any details. He was not pinning his hopes on the aviary.

'I have been possibly the most prominent critic of this system,' he said. He thought it was ludicrous to believe that Parsi bodies could be certifiably diclofenac-free.

'Seventy-two hours before the person dies, he is not supposed to be taking diclofenac,' he said. 'Now who the hell knows when the person is going to die? You know, I was driving down from Pune just now, and I may have a heart attack, and I may be admitted to a hospital. You think the doctor is going to be asking me all these questions? I will not be in a position to answer any questions and he's going to administer life-saving drugs. I mean, it's ridiculous. It's preposterous.'

Early in 2014, a letter signed by Dr Jehanbux A. Chichgar and eighteen other Parsi doctors was sent to the Bombay Parsi Punchayet, absolving themselves of any responsibility regarding a Parsi aviary. There were 1,051 combinations of diclofenac, they wrote. There were studies emerging that indicated it might not be just diclofenac that was toxic to the vultures, but other NSAIDs as well. As little as 0.004 micrograms of diclofenac in a corpse could cause vulture mortality.

Homi's grandchild entered the room quietly as we talked. He circled Homi's chair, played with a small ornament on the table, and disappeared again. I asked Homi what he would like to be done with his own body when the time came.

'I would prefer to be put in the towers,' he said, 'and by the time I close my eyes, I think the system will work very effectively, very efficiently. Let us hope.' The light was gone from the sky outside the window.

'I think it should be to each his own,' he said. 'We are living in a democracy. If a man wants to be cremated, so be it. If a man wants to be buried, so be it.'

Homi, like Dhan Baria and others, are ready to allow burial or cremation into the Parsi parameters of ritual, eager to help their religion find a way to adjust to the vultures' absence.

There were, technically, other options. They could move

from the archaic to the cutting-edge of green technology and use gasification, a process that involves high heat that never enters direct contact with the body, or promession, a technique catching on in Sweden that uses liquid nitrogen to deep-freeze a body before vibrating it into a fine dust. One prominent bird breeder suggested that Parsis try using insects. All these ideas seemed too strange. And dhokmanasheeni dictates exposure to sunlight as well as predators; so all recommendations so far have been met with dismissal. The Parsis continue to point their solar collectors at their dead and hope for the best.

'They're gone. They're gone. They're gone,' said Dr Asad Rahmani, director of BNHS, when I first sat down with him at the BNHS headquarters in Mumbai.[48] While Vibhu remained open to the possibility of a successful vulture aviary at the Doongerwadi, Asad was more sceptical.

'If one Parsi comes up with diclofenac in the body, half the [vulture] population will disappear. We can't afford to lose them,' Asad told me. His large desk was topped with a sheet of glass. Plastered beneath the sheet were slips of paper and photographs including a quote from American diplomat Edward John Phelps: 'The man who makes no mistakes does not usually make anything.'

Diclofenac use on humans began about forty years ago, he said. Veterinary use began in the early 1990s, yet the first vulture declines were noted as early as the 1970s, increasing in the 1980s and becoming devastating in the 1990s. Could the Parsis, who depended upon the vultures in the most intimate of ways, possibly have begun the collapse? Asad and others said it was certainly possible.

'We have to win,' he said of BNHS's reintroduction efforts. 'We don't want the vultures to be only in big cages. There is only one problem, there is only one solution.' A slew of threats—

including habitat loss, hunting, diseases erupting in a warming world, toxic exposure, poaching and more—endanger so many of our world's creatures. Vultures dying from diclofenac poisoning is, on the other hand, just one isolated problem. 'If you remove poison from the food,' he said, 'they will do well. They will not come back to forty crore population, but at least they will be safe.'

'What do you hope to see down the road?' I asked him. A laugh erupted from his large frame.

'Vultures flying all over India! Seeing 150 vultures on a cattle carcass finishing the carcass quickly, like we used to see twenty to thirty years ago.' Asad suddenly became serious and wistful.

'Who thought that the vultures would disappear so quickly? I never thought ...,' and his sentence trailed off. 'This is a very very personal sadness to me. There were so many that the younger generation cannot even imagine how many vultures used to be there. They were everywhere. You go from Aligarh to New Delhi, and you would find vultures flying all around. And I remember when I was a child, in winter, when we used to go and lie down in the afternoon, you looked up and saw huge flocks of vultures would come and pass.'

He was a man filled with regret for losing something he'd neglected to appreciate. 'At that time, the digital camera was not there. It was so expensive. Every click is important. So who will spend clicking vultures? They were so common; why should we take a picture of a vulture? Just to imagine that they would become rare was impossible. Now there are dogs. They eat anything, live and dead. There are dogs on the ground but the skies are empty.'

He broadened the conversation out and spoke of the dismal state of conservation in today's India.

'Conservation money is only naam ke vaaste,' he said. 'It is not real; it is only for the name's sake. Our hands are so badly tied up in India. No other country will have so many restrictions on wildlife research. We can't do anything. For telemetry, for taking samples—everything is difficult.' Support from the public is slim as well. BNHS, the largest conservation organization in India, in existence for well over a century, has only 5,000 members.

Conservation was neglected unless it was for tigers. 'Who is interested in vulture conservation?' Asad asked. 'In India, the highest priority is tigers.' He paused, then corrected himself. 'The *only* priority is tigers. Seventy per cent of the Ministry of Environment and Forests (MOEF) budget is for tigers. We are not saying reduce that 70 per cent. Increase the others.'

The annual budget for the vulture programme at all the BNHS centres is roughly Rs 2 crore. Tigers, in comparison, are allocated over Rs 171 crore by the Indian government.[49]

'In another ten years, we will see a lot of extinctions in India. Less than 150 hanguls are left,' he said, of the Kashmiri stag with the five-tined antlers. Only twenty-five wild Asiatic buffalo remain in the Bastar area of Chhattisgarh. Fewer than 300 great Indian bustard roam through grasslands.[50]

I realized as I spoke with Asad that I had come to India looking for an eco-catastrophe. Though the vulture is the unloveliest of creatures, though few cared for them while they were here nor mind that they are missing, their absence has left a void. There is a physical abyss that seems to be filling with feral dogs, and a spiritual vacuum that is forcing questions of adaption for the most orthodox of India's Parsis.

Yet the situation didn't feel apocalyptic to me; it just felt empty. Maybe all of us, whether guided by God or by science, secretly want to be the ones living in the end times, as though it bestows some epic importance upon our lives. What if there was no mass annihilation, but instead a slow accumulation of minor deaths that no one notices? The vultures' disappearance is catastrophic, yes, but our ability to adapt is stunning. No matter how many species get wiped from the earth in the wake of progress, the living go on. Species that disappear are erased from the bio-narrative of the planet and forgotten. Like Asad said, there are young adults in India who have never seen a vulture. As the vultures became functionally extinct in under a decade, India's ecology has shifted and the habits of her people, whether farmers or skinners or Parsi priests, have too.

'I wish we were not so flexible,' Asad said. 'We're like pests. We can live in every type of environment, from Alaska to Namibia. We're omnivorous. We're like cockroaches, rats and bandicoots. You have seen the slums,' he continued without reservation. 'Look at the horrible conditions we can live under and still have reproductive success.'

He leaned back in his chair and the continuous insistent blast of car horns filled the space between his words.

'No other species has such a huge tolerance. For the earth, that is the unfortunate part. If we had a very narrow tolerance level of pollution and food and all these things, maybe we would take more care of the earth.'

Asad spent his days supporting the multiple efforts under way to save the vultures, but his words betrayed his sense of hope.

'Nature is dynamic. It is not the balance of nature; it is the dynamism of nature. It is a terrible thing to think of a world without the vulture, but what does it matter? Nature settles itself.'

BASEMENT STRATEGIES

In late 2013, the skies were still empty, and the third meeting of SAVE (Saving Asia's Vultures from Extinction), a consortium of regional and international organizations that were coordinating conservation, campaigning and fund-raising efforts, attracted enough people to fill a basement in West Bengal. Fifty delegates from six countries were present. India and Pakistan were working together and had partnered with Nepal and Bangladesh among others. Asad Rahmani arrived late from Nagaland, where a rare success story of amur falcon conservation was unfolding. Vibhu Pakash and his wife Nikita came from Pinjore. Percy Avari, a Parsi veterinarian from Mumbai, and a dozen other people I'd spoken to over email and phone in my years reporting on vultures were there. Over the course of three gorgeous days in November 2013, we sat in a bare room under a ceiling that scraped the tops of our heads and strategized about how *Gyps* vultures could be made abundant again.[51]

There was hopeful news. Multiple breeding centres were up and running: the original and largest at Pinjore; a second in Rani, Assam; and a third, Rajabhatkhwa, tucked into a corner of the nearby Buxa reserve. Back in the Pakistani forest of Changa Manga, where some of the first birds were collected and tested in 2000, there was another breeding centre, and there was a fifth in Nepal. Researchers were continuously honing breeding methods including artificial incubation and double clutching, ever increasing the number of babies they could coax from each mating pair.

Altogether there were a few hundred vultures at these centres. Ninety vulture chicks had made it to their fledgling flight, though no birds, young or old, had been released from captivity. The Indian government's Central Zoo Authority had set up five additional breeding centres in zoos around India. The rapid decline of the last decade had stabilized and populations in a few places had even gone up, though numbers were still gravely low.[52] Diclofenac use was definitely dropping, but it was still present in livestock carcasses in dangerous levels.

In 2006, the German company Boehringer Ingelheim developed a vulture-safe alternative called meloxicam. At the prompting of BNHS and other international conservation organizations, it generously released its license so that India could manufacture meloxicam without restrictions. Between 2006 and 2011, the governments of Bangladesh, India, Nepal and Pakistan all officially banned veterinary formulations of diclofenac, though unofficially many veterinary stores still carried it under the counter. Human doses of diclofenac remained legal and readily available, and as long it was for sale, farmers and vets would likely continue to use it.

The farmers' resistance to meloxicam was due to the fact that it was too acidic, causing discomfort to the animal when injected. It was also slow to take effect and more costly than diclofenac.[53] Government subsidies were attempting to bring parity to the prices.

'It's not about meloxicam being more expensive,' Vibhu told me later.[54] Cattle and water buffalo can be worth a lot of money; they are major investments for a farmer, 'so if it falls sick, it will get the kind of treatment that a human would get.' When I'm in pain, the farmers think, I take diclofenac, not meloxicam. So why not do the same for my cows?

Good news was that influential players such as Amul Dairy in Gujarat, the largest cooperative dairy in India, were adopting meloxicam on a large scale. Guided by the Mohamed bin Zayed Species Conservation Fund, the dairy, which has over 2,00,000 farmers and 150 staff veterinarians caring for their dairy cows, administers only meloxicam. The food and drug authority in Gujarat showed quick support for meloxicam when approached by conservation organizations, immediately issuing a circular to all district-level drug inspectors for stringent regulation over veterinary use of human diclofenac.[55]

There was also pressure on pharmaceutical companies to eliminate one ounce multidose vials of diclofenac, which make it extremely easy to supply the black market, since one of those vials is perfectly sized to shoot up a couple of cows.[56] Smaller vials would make that practice inconvenient. Three companies have already agreed, including the major supplier Novartis.[57]

Across northern India and into Nepal, seven safe zones for vultures were established: four hugged the border between the two countries, one lay in the eastern state of Assam and two were in Gujarat. In these zones, efforts to completely eradicate diclofenac are concentrated in a 60-mile-radius area, covering more than 11,000 square miles, though vultures are known to roam much farther than that.[58] The conference was hopeful that the first captive birds could be released within one of these zones in 2016 or 2017.

Even as that hope hung in the air, there were disquieting reports from researchers who had found evidence that other birds were being injured by NSAIDs.

Dr Anil K. Sharma, principal scientist in charge of the Centre for Wildlife Indian Veterinary Research Institute in Izalnagar, Uttar Pradesh, had done the pathology on two steppe eagles (*Aquila nipalensis*) found dead at the Jorbeer dump in Bikaner. These eagles, closely related to the golden eagle, showed the same clinical signs of kidney failure seen in *Gyps* vultures. One of the birds had diclofenac residue in its tissues.[59] It was the first time the drug had been detected in a non-*Gyps* species.

Toby Galligan, a young Australian conservation scientist with the RSPB, reported surveys showing that Indian populations of Egyptian (*Neophron percnopterus*) and red-headed (*Sarcogyps calvus*) vultures had declined with similar timing and scale as *Gyps* vultures.[60] (Old World vultures of Asia and Europe and New World vultures of the Americas, though they appear quite similar, are actually genetically distinct.) Researchers were developing a vulture kidney cell line that might help reveal why diclofenac is so toxic to some birds and not others.

Scientists were also wrestling with new research proving that other drugs were also killing vultures and other birds. Diclofenac is just one of a long list of NSAIDs with unpronounceable names. Meloxicam had proven safe so far,[61] but could ibuprofen, ketoprofen, oxyphen, nimesulide, piroxicam, mefenamic acid, phenylbutazone and paracetamol prove to be as lethal as diclofenac?[62] Six dead birds were found with high levels of nimesulide, a veterinary painkiller that was growing in popularity, and four of them had visceral gout.[63]

There was glum news that Spain and Italy, home to some of Europe's largest populations of vultures and eagles, had recently approved veterinary use of diclofenac.[64] For vulture conservation to have any hope of success, disparate government ministries and non-governmental organizations will have to collaborate, along with veterinarians, pharmacists, drug-marketing professionals, cattle owners, milk cooperative societies, gaushalas and panjrapoles (old-age homes for cattle), animal welfare organizations, forest departments, youth, elected representatives, media and local non-profit organizations.

And safe zones for vultures have to work. They have to be at least close to 100 per cent diclofenac-free before they can release a single bird, and it has to happen soon.

'We cannot,' Vibhu said at one point, 'keep the birds forever.'

At the end of the day, some of the group retreated to a windowless dining room to continue their singularly focused conversations about vultures. We drank Carlsberg Elephant beer from tall green bottles and mysteriously green-tinged gin and tonics, accompanied by French fries that arrived in slow order.

'I have faith this is going to work,' Campbell Murn, scientific officer at the Hawk Conservancy Trust in the United Kingdom said. 'I can imagine myself down the road, an old man with my flannel shirt and a mug of hot broth in my hands, looking back and thinking we were the ones who did it.'

Daniel Wilcox, based in Cambodia with BirdLife

International, laughed and shook his head. 'I'm not so optimistic.'

'One thing this has done is to make us not take anything for granted,' Hem Sagar, the Nepal coordinator for the Zoological Society of London, said. 'We knew nothing about the vultures. They were trash birds which we ignored because they were everywhere.'

Everyone nodded in assent.

After the SAVE meeting in West Bengal, I returned to the Pinjore breeding centre in Haryana with Vibhu and Nikita and a few others.[65] It had been four years since I was there last and the centre had grown, rechristened the Jatayu Conservation Breeding Centre. Vibhu was working in his office as Nikita led me off to show me the changes. Her wonder at the vultures she worked with hadn't flagged in the four years since I'd first met her. Across the 5 acres, there were more aviaries, more birds. In 2009, the centre had 123 vultures. By 2013, there were about fifty additional birds and two drops of blood from each were tucked away in a freezer as part of the Frozen Ark project,[66] which archives the DNA of endangered animals across the planet.

Nikita and I sat in the dark room of a new interpretive centre, large enough for school groups and others to visit, watching the same real-time video feed from the aviaries that we had watched four years ago, but this time the screen covered an entire wall. A close-up of a long-billed vulture showed us the slender creamy talons of its nails, its vanilla-hued tertiary feathers, the darker secondaries and chocolate brown primaries.

Later, Nikita and I circled back to the main buildings and another new addition to the centre: an incubation room being readied for the first eggs of the season, which would arrive in nests at any moment. It was a sunny warm afternoon, and I helped Nikita and two staff men wipe the yellow and black incubators down after they'd been sterilized, eleven little shoebox-sized machines that could each cradle and rock four vultures' eggs into a chick.

The months ahead would be the busiest of the year for the staff. Eggs needed to be placed in the incubators. Temperature and humidity needed to be monitored around the clock. And everything was to be noted in a fat school ledger and entered into a computer database.

Nikita held a dummy egg up to her ear to show me how she would listen to the chick pipping through the eggshell. A day or two later, the chick would break through. Nikita cupped her small palms together, in sacerdotal offering, to show me what the shape of the chick would be. After four months of painstaking care, the young vultures would be introduced to the aviaries to join the adults and other young that had been left to be raised by their parents. And then all would resume the wait for a world that is safe for them beyond the confines of concrete walls. It was Life Row.

Later, I joined Vibhu in his office to ask what he thought about the cheery newspaper articles that appear occasionally, all with a variant on the title, 'Vultures Back!'

'That talk of vulture populations coming back is rubbish!' he said. Since the early 1990s, Vibhu and others have been driving road transects across northern and central India, racking up nearly 10,000 miles each time, counting birds. 'We see the raw data. We know the situation is grim,' he said.

But the reason behind the newspaper stories was positive. Earlier, no one was looking for the vultures, Vibhu said. Now they are. The crisis is drawing a young generation of Indians to conservation. This gives him hope.

'The country has a lot of problems,' he admitted. 'Of course it is frustrating that actions have not been taken very fast, but I live in this country and, so far as vultures are concerned, the country is doing very well. All these NSAIDs are very good for cattle and ours is a country where dairy farming is a major occupation. Government authorities do listen and do take action. But we must understand that things go slow in this country.'

His leg jiggled under his desk. 'The good thing is that there is so much awareness of vultures now. Everywhere you go, wildlifers talk of vultures only. Action is happening. I'm quite positive about it.'

I asked what he hoped to see come of all his work.

He paused. 'If they're able to survive and breed in the wild, that will be a very good test for us,' he said. 'But releasing birds will be the most exciting part of it.'

Years after I held the sick vulture back in Rajasthan, I can still feel its heft. I can imagine how it would feel in Vibhu's hands, standing at the heart of a safe zone for vultures, holding it, and then the moment of release. And I can imagine the weight that will linger even when his hands are empty, the vulture aloft.

I kept a journal on a trip to India when I was ten years old. In it, I dutifully noted facts I was learning about my father's homeland. 'There are,' I wrote in a careful penmanship, 'one hundred million forms of god.' I didn't know then that there was a vulture god. I didn't know to look up and still see them flying.

When I had stood at the edge of the Jorbeer carcass dump with Jitu Solanki, he had spoken of mythology and holy things. He rhapsodized about the Indian's intimate connection with the natural world as we tried to stand upwind from the rot. The young naturalist had gestured to a nearby tree, where griffons topped the branches. 'This tree that you are seeing here, this is the state tree of Rajasthan. We call it khejari tree, and we say it is a holy tree.' *Prosopis cineraria* can tap deep into hidden aquifers, drawing life out of a seemingly lifeless desert.

'So everything you can use from this tree, but you never cut the stem; you always cut just the small parts. The local people believe if you cut this tree, you will go to hell,' he told me, and smiled.

'Our forefathers were so clever when they were making their religious rules,' he said. 'If the tree is useful, people will cut it down, but if the priests say it is holy, people will let it grow.'

These beliefs have kept Hindus from killing cows and dogs, but the faith of nine hundred million Hindus in their god Jatayu is not saving the vultures. The Parsis who have lost a part of their tradition are adapting.

The thing about mythology is it outlives the circumstances on which it was based. Lord Vishnu will still climb upon his bird-man mount Garuda—the mythical creature from which all birds of prey are descended according to the Mahabharata—even if not a single vulture flies free over South Asia.

And the thing about extinction is that it is so very quiet amid the decibels of daily life. My chase after the vanishing vultures across South Asia felt like the pursuit of a whisper. This is what extinction looks like. It is as silent as a raptor in flight, coasting on broad wings, filling the liminal space between heaven and earth. The millions of vultures that have died fought no valiant battle except against a modern chemical world they cannot comprehend.

But the vanishing have defenders, not least those few dozen people who came from all over South Asia and far beyond to sit in a basement. It actually takes very few people to enact change. 'A small group of thoughtful people could change the world,' American anthropologist Margaret Mead famously figured. Scientists like Munir and Patrick who are willing to stare up at cliff faces for weeks on end, year after year. Biologists like

Vibhu and Nikita preparing for a new breeding season. An official like Dr R.D. Jakati of the Haryana Forest Department welcoming BNHS to set up a breeding centre. A businessman like Balendu Singh willing to open his hotel doors to a stinky, sick vulture.

All it takes is for someone, anyone, to say yes instead of no.

5

AKASHA

ETHER

Why shall I wait for someone else? Why shall I be looking to the government, to the army, that they would help us? Why don't I raise my voice? Why don't we speak up for our rights?[1]

—Malala Yousafzai

WELCOME TO BIHAR

Pinki Kumari didn't flinch when her interviewers showed her images of human genitalia. She was twenty-three years old, and utterly bored with her job preparing accounts in a CPA's office. A friend had told her that a group called Pathfinder International, a global organization that educates the poor on reproductive health in more than twenty countries, was hiring. She applied, having no idea of what to expect. During her interview, a charismatic and lanky project manager named Binod Singh, who had talked with dozens of women like Pinki, handed her one of the organization's educational flip books with the explicit photographs.

As a native of Bihar, Pinki had spent part of her life in a village, part in the capital city of Patna. Her mother was a nurse who'd brought home all sorts of health-related literature. Still, Pinki had never seen such graphic images. But she wanted the job, so she imagined she was looking at them as a doctor would.

'I thought, "If I had to do an operation, it would be this same body that would be there, so I have no issue with this."' She remained nonchalant, but Binod pressed her.

'Boys will ask you questions and the community will abuse

you,' he said. 'How will you handle it?' She was unflappable. In her school days, she had dreamed of being a doctor, so she figured this was a way to be a social doctor, one who could educate people on health issues. She told Binod as much.

A week later, she was hired, and after a month of training Pinki began travelling from village to village to teach adolescents the fundamentals of reproductive and sexual health. This knowledge—what menstruation is, how to use birth control, and why delaying marriage and staying in school is in a kid's best interests—was both subversive and profound.

'In eleven years, I must have interviewed 2,000 people!' Binod recalled, 'I tell you, I was really getting bored. But right away with Pinki I saw a spark and honesty.'

It was seven years after that interview, and Pinki, Binod and I were travelling from Patna to the town of Bodhgaya to meet some of the unmarried adolescents Pathfinder had trained.[2] Sitting next to me, Pinki was a composed presence in a purple salwar kameez, wearing thin-framed glasses and simple gold hoop earrings. We were travelling south, away from the latitudinal course of the river Ganga, which bisects Bihar at its midsection. Winter was approaching, and the people we passed were swaddled with scarves in the early morning cool. We left the heart of the city behind, but evidence of its expansion lingered. Buildings were under construction everywhere. Men with shovels filled plastic bags with sand and others carried them on their heads up half-built flights of stairs. Everywhere there were structures with pieces of rebar jutting out into the sky.

Farther from the city, rice paddies appeared. Lotus flowers, delicate vestiges of Bihar's Buddhist roots, poked out of roadside ditches. Singharha, water chestnuts, choked the wetlands. Kids scavenged for recyclables in mounds of makeshift landfills, and kiln chimneys interrupted the flat landscape, perched on pillars of red earth, encircled by piles of bricks. A black-shouldered kite hovered in the haze over the plain. Colour came in quick splashes—fleeting views of a brightly painted truck or an Amul Macho Innerwear advertisement blanketing the side of a shop.

I had come to Bihar to find out what it means to be a young woman in today's India. I was more familiar with the India of the south, where my father's family lived, where women and men share the streets in equal measure. Bihar was different. I felt the weight of purdah—the practice of keeping women out of view, based on the Persian word for veil or curtain, wherever I travelled (and not just among Muslims). In Bihar, especially in Patna, men dominated public spaces. I glimpsed women and girls in small doses, occasionally socializing in coffee shops near the university, walking the perimeters of their rooftops for exercise, in pairs appraising fabric in corner shops. For women, to step outside was to invite risk.

Bihar is a beleaguered state with a notorious reputation for violence. It scrapes the bottom of nearly every socio-economic ranking there is: per capita income, sex ratio, and rates of illiteracy, malnourishment and infant mortality.[3] Economic and infrastructural improvements picked up after 2005 when Nitish Kumar, son of a freedom fighter and a socialist, became chief minister, but even after nearly a decade of Indian economic growth, more than half of Biharis still lived in poverty. If factors beyond income were considered (using the Multidimensional Poverty Index), that figure came closer to 80 per cent.[4]

Population growth, though most rapid in Bihar and its neighbouring states of Madhya Pradesh, Rajasthan and Uttar Pradesh—collectively known as BIMARU—is a problem for the entire nation. In 2011, the seven-billionth human wailed its way into the world, a miracle emerging from the ether, life materializing from a web spun of spiralling DNA. There is a fine chance that this child was born in India. In the spring of the same year, India's population topped 1.2 billion—triple what it was in 1947.

I have witnessed this growth in my own family. In 1921, my grandparents were married in a small south Indian village in the present-day state of Tamil Nadu. My grandmother was just a girl of ten, my grandfather a boy of sixteen. After days of celebration, they returned to their respective homes and didn't move in together as man and wife until she'd reached the tender age of sixteen. This practice, commonly referred to as a 'second marriage', is still common for many young couples in rural India today.

My father is the middle of nine siblings. Eight of nine survived, growing up to find spouses and raise children of their own. As of this writing, I have nineteen first cousins in a family of eighty-two. While fertility rates in Russia, Europe and Canada plummet, India's rate remains high. At its current pace, India won't achieve population stabilization—2.1 children per mother—until 2060.[5] By then, it could contain 1.62 billion people.[6]

This rapid rise in population impacts every aspect of life, from availability of food and water to job opportunities. 'The US is, in area, three times larger than India,' as one Bihar government official told me, 'and its population is one-fourth of India's. The pressure on land here is more than ten times what it is there.'[7]

And that pressure is even greater in Bihar, already one of the most densely populated states in India, and getting more crowded every year. Bihar has one of the nation's highest fertility rates and some of its youngest brides. It is a dismal place to be a woman. According to the World Health Organization, for every 1,00,000 mothers who give birth in India, nearly 200 die—more than seven times the US figure.[8] But states with better health care and education, like Kerala in the south, have a third of Bihar's maternal deaths.[9] Though it is illegal, nearly 70 per cent of Bihari girls marry before their eighteenth birthdays,[10] and well over half have their first child before they are nineteen.[11] Of all babies born in Bihar, nearly 5 per cent die within the first year, though at least that figure is decreasing.[12]

Women don't fare much better in India as a whole. In 2012, a Thomas Reuters Foundation report measuring the threat of physical and sexual violence, quality of health services, level of political expression, and access to property and land rights, found that of all the G20 countries, India was rated the worst for women, followed by Saudi Arabia.[13]

Girls across South Asia are fighting for the right to an education and the right to be physically safe at home and in public. My stay in Bihar was bookended by two events in South Asia that made international news. On 9 October 2012, less than a month before I first met Pinki at Pathfinder, fifteen-year-old Pakistani Malala Yousafzai was shot in the head by the Taliban in response to her outspoken efforts to keep girls in school. Two months later, a twenty-three-year-old female physiotherapy student named Jyoti Singh Pandey and her male friend, after watching *The Life of Pi* one evening, boarded a private New Delhi bus they thought was offering them a ride home. Behind the bus's darkened windows, six men violently gang-raped the young woman with a metal bar. They left her and her beaten companion on the side of a road. The man recovered from the attack. The young woman, whose insides had been mutilated beyond repair, died thirteen days later.

This rape made major headlines around the world, portraying a side of India both misogynistic and savage. Enraged protesters took to the streets in New Delhi, demanding punishment and reform. Long before this attack, I had been reading about sexual violence in the local papers. Every day, women and girls were assaulted on their way home from visiting relatives, or lured by a stranger posing as a family friend, or attacked when they walked to a field for relief because they had no toilet. Two five-year-old girls were brutalized in separate occurrences. One died. A seventeen-year-old rape victim in Punjab committed suicide after police presented her with two options: accept a financial settlement or marry one of her rapists. A young photojournalist and her male colleague were attacked at an old mill in Mumbai—a site that serial rapists prowled, calling each other whenever a 'beautiful deer' had been spotted. Another woman had been raped in the same place and had been subjected to a medically debunked 'two-finger' hymen test by doctors to determine whether or not she was lying.[14] The chief minister of West Bengal—a woman—dismissed the highly publicized Park Street rape case in Kolkata as a shajano ghatana, a 'fabricated incident'[15] made up by the victim in order to 'make her government look bad'.[16]

The twin events of Malala's shooting in Pakistan and the New Delhi gang rape ignited protests in South Asia and around the world. Working with unprecedented speed, the Indian Parliament responded to the New Delhi gang rape by expanding the legal definition of rape and criminalizing sexual harassment, voyeurism and stalking—acts still widely dismissed as 'Eve-teasing' in India. Marital rape is still not considered a criminal offence.

The law was a leap, yet grave concerns remain. Victims often find the police as threatening as their attackers, and the judicial system is notoriously torpid.[17] It is not unusual for ten years to pass before a judge reviews a rape case. After the high-profile New Delhi incident, fast-track trials were implemented, and four of the rapists were sentenced to death less than a year after the attack,[18] but it is believed that many rapes remain unreported—the vast majority by most estimates.[19] According to the National Crime Records Bureau (NCRB), only 24,206 rape cases were registered in India in 2011.[20] And the sexual violence is by no means directed solely towards females.[21]

The greatest concentration of women I witnessed in any one place in Patna was at a bend in Boring-Patliputra Road. Inside the Pathfinder office, not far from the World Class Infertility Clinic, bands of eager young women, along with a few male colleagues, swept through the door and past the table where I first met Pinki in a golden salwar and Binod in a bold purple shirt.

In the office where we all first met and on our trip to Bodhgaya, I could see the devotion Binod felt towards Pinki. It seemed to have stemmed from that first interview he had with her and to have grown in the tumultuous years of her life that followed. He had supported her personally and professionally, but it was Pinki who forged her own life.

Pinki was one possible vision of India's future woman. She was not the more commonly portrayed girl who slipped on jeans and fled to the city to work in a call centre. She had forsaken neither her religion nor her parents, but she was redefining her identity as an Indian woman according to her own parameters from the time she was sixteen.

Girls in India are now demanding the rights that have been unquestionably afforded their male counterparts. While some men support them, many others are putting up fierce resistance, sometimes subtle, sometimes violent.

I was drawn to Bihar to explore population, but it was power I found at play. Pathfinder was bringing health and education, power-masked, to a new generation. And power is political. My new question became: Is it finite? Is one person's gain necessarily another person's loss? Is power as limited as a cup of tea? If one person drinks, is another left thirsty? Because girls in India are parched. For knowledge. For their chance. For the right to control their lives, to choose their futures. For the right to not be attacked by gangs of men or uncles with access.

'Girls are not waiting for any opportunity. They are just grabbing the opportunity from the male,' Pinki told me at one point in the car, gazing out of the window at the landscape of rural Bihar where she had been raised. Her experience had introduced her to a generation of girls ready to fight for their rights to knowledge and safety. She had seen girls who worked harder in school than the boys. Girls who went to extreme lengths to find their own place in the world.

'They come from a deprived community,' Pinki had told me, 'where they never talk about themselves, where they never think about themselves, their life, what they have to do. Once they have a platform, they start thinking, they start to dream. Yes, I can be a leader. Yes, I can speak for myself.'

Malala, who survived the attempt on her life and won the Nobel Peace Prize in 2014, wrote: 'I dreamt of a country where education would prevail.' Dreaming can be transformative. It can also be dangerous.

A GOLDEN CHANCE

Beneath the troubles of modern-day Bihar lies a glorious past. The entire region is one of the oldest inhabited areas on earth, layered with eras of religious, political and intellectual plenty. In the sixth century BCE, Siddhartha Gautama—a prince who sought to release himself from the suffering of the world—found enlightenment under the tendril-tipped leaves of a bodhi tree in Bodhgaya and thus became known as the Buddha, the One Who Woke Up. Around the same time, Jainism emerged with the birth of a man named Mahavira, born among the Rajgir Hills of central Bihar. These hills are south of modern-day Patna, a city earlier known as Pataliputra, once the capital of the Mauryan Empire, which ruled Central Asia from 321–185 BCE[22] under the hand of the powerful and enlightened

King Ashoka, who spread far and wide the Buddha's teachings. The Pataliputra region was also home to two international universities: Nalanda University, which once taught ten thousand students, and Vikramshila University, which flourished between the fifth and twelfth centuries AD.[23] Pataliputra had universal and free medical services.[24] In 1918, Gandhi's Satyagraha movement against British colonial rule was born in Bihar when Gandhi inspired a non-violent farmer revolt protesting the forced production of indigo for the British.

But as the rest of India developed in the second half of the twentieth century, Bihar stagnated and regressed, crippled by a complicated amalgam of poor policy decisions, unequal land distribution and underdeveloped infrastructure, which doomed its citizens to a semi-feudal existence.[25]

Still, the small town of Bodhgaya is a travel destination for Buddhists around the world who come to visit the holy sites, especially the Mahabodhi temple and the sacred holy fig next to it, where the devout and the curious sit under the canopy of a tree allegedly descended from the one the Buddha-to-be meditated under for twenty-nine days.

On the far side of town, miles from the sacred tree, Pinki, Binod and I arrived at the Sujata Kuti stupa. Named after the maiden who brought kheer to the Buddha soon after his enlightenment, the site had no ticket booths or metal detectors like the Mahabodhi temple in town. Tucked behind the nondescript dome of stone that marked the spot was Bakraur, a village ringed by fields of ripening cabbage and eggplant. Across from Sujata Kuti stupa was a small brick one-room schoolhouse. Inside, there was just enough room for us to sit in a circle on a thin, damp rug with two dozen teenagers who had come from Bakraur and villages beyond to meet us.[26]

The girls outnumbered the boys. All had gone through a three-day training session that Pathfinder offered on basic reproductive and sexual health, part of its PRACHAR

(Producing Change in Reproductive Behavior) project. Prachar means 'to promote' in Hindi. Some of the boys and girls were also involved with another Pathfinder programme, Jagriti, which means 'awakening', that aims to end child marriage, a practice that was banned in 2006 but remains commonplace.

The teens were of mixed castes, though many shared the common last name of Kumar/Kumari, the same as Pinki's. The enclosed space was illuminated by dim daylight that seeped in through half-shuttered windows and lit their young faces and the faded blue plaster walls. Each person shared a story. By most accounts, the three-day Pathfinder training initiative was transformative. The kids used the term jagriti frequently as they spoke of their own awakenings. They seemed enraptured with Pinki and Binod—and why wouldn't they be? Binod was charming and informative. He teased and joked. Pinki was, for them, a rare example of a working woman, happy and passionate and brazenly (or bafflingly) without either a husband or a child at thirty years old, which is nearly unheard of in rural India.

Nobody else had ever told these teens that they had the right to make decisions about their own lives. No one had offered them practical information about topics ordinarily considered taboo, and thus made all the more alluring.

'It is up to you to decide when you should marry,' Pinki and Binod told their students. 'Wait until at least the legal age of eighteen, and stay in school as long as you can.' 'You and your spouse can decide how many children to have and when to have them.' 'It is your right to control your fertility.'

The class also learned about basic contraception. That it's impossible for an intrauterine device (IUD) to end up in your brain or your windpipe. That it is the father's biological contribution that determines a child's gender with a tiny thing called a chromosome, so if anyone is to be blamed for the usually conceived unfortunate arrival of a girl child, it is the father. The girls loved this fact and repeated it often.

Pathfinder kept meticulous track of its trainings and the outcomes, part of a global trend to quantify the effectiveness of social-service programmes in order to make the work more effective and keep the Western funding flowing. Pathfinder reported that in the populations of unwed adolescents whom it had trained, the average age of marriage had gone up 2.6 years, while the arrival of firstborns had been postponed by 1.5 years compared with control groups.[27] By delaying marriage, Pathfinder enabled girls to continue their schooling while their bodies matured. Then couples could find their own way with each other and into adulthood before taking on the demanding role of parents.

These are not just feel-good platitudes. There is life-saving science behind Pathfinder's work and India's prohibition of marriage before the age of eighteen. For girls between the ages of fourteen and nineteen, pregnancy-related deaths are the most common cause of mortality.[28] These young mothers-to-be are twice as likely to die in childbirth as those in the 20–24 age bracket.[29] According to the United Nations Fund for Population Activities (UNFPA), nearly 63,000 Indian women die annually—that's one every eight minutes—due to causes related to pregnancy and childbirth.[30] This figure accounts for 18 per cent of estimated global maternal deaths, a leading cause of adult female mortality worldwide.[31]

For the young brides who survive, motherhood may garner them greater respect in a culture that so values marriage and children, but at what cost? A Guttmacher Institute study found that an Indian bride who married before she was eighteen was less likely to have any say about the man she married. He was likely to be older. And she was less likely to feel she had the right to speak back to elders or anyone else with whom she disagreed. Younger brides faced more physical and sexual violence, often from their own husbands, and were more likely to think such violence justified. They were less likely to use contraception, they had less control over their finances, and they found it harder to move through the world (to a store, say, or to a friend's house) without permission. They were less likely to see a film with their husband or take a trip to visit their natal home. They were also less likely to have a first child within a health-care facility, and more likely to miscarry.[32]

Global health expert Rob Stephenson found a direct relationship between domestic violence and the ability of a woman to 'achieve her fertility intentions', with abused women less likely to use contraception and thus more likely to have unwanted children.[33] Other studies showed that this lack of autonomy is related to a deficiency in the use of pregnancy-related care, which in turn leads to higher maternal mortality rates.[34] And so the cycles of poverty and poor health are carried on to the next generation.

In the face of this grim data, Pathfinder's mission is modest: to encourage women to delay marriage and space the timing of their children, instead of marrying early, immediately giving birth multiple times, and then getting sterilized. (Currently, sterilization accounts for 85 per cent of India's female contraception.)[35]

Controlling conception is a contentious topic in India thanks to the lingering legacy of the Emergency, which Prime Minister Indira Gandhi declared on 25 June 1975. With this declaration, India descended into a twenty-one-month-long period in which Gandhi ruled as a despot, suspending fundamental rights, imposing censorship and jailing protestors. Her son Sanjay took on population control as a personal pet project during this period, and though he held no official political title, he set up an extensive programme to sterilize men. According to journalist Mara Hvistendahl, this was the extreme outcome of a twisted amalgam of ideas originally introduced to India after World War II by Western advisors from the Population Council, UNFPA, Ford Foundation and World Bank. During just one year of the Emergency, Hvistendahl wrote, 6.2 million Indian men were sterilized, fifteen times the number of people sterilized by the Nazis.[36] Nearly forty years later, sterilization is the dominant form of birth control but almost exclusively it is women who get this done via tubal ligations. It seems that no one, male or female, wants to revisit the atrocities of the makeshift vasectomy camps from that troubled era.

'Why do we give this three-day training?' Binod asked the teens. 'This is rare that men and women are sitting here together. Previously, women would be separate. Did we make a mistake?' He held his hand up, long fingers stretched out in a questioning gesture. 'Oh God, oh mother!' he cried, 'Everything's going to hell!' He laughed and the kids laughed with him. Then he paused, and said firmly, pointing one of those long fingers at them, 'Don't think this way.'

'We should be fearless,' Pinki added.

The kids broke into applause. The girls clapped loudest and longest.

None of the kids in the room were married, though one girl was engaged. Binod instructed the others not to tease her. One boy named Papu Kumar, with big ears and high cheekbones, had three younger sisters. He declared that he had decided to wait until twenty to get married. He was encouraging his sisters to do so as well. But Sanoj, a forlorn sixteen-year-old boy in a frayed white baseball cap, spoke up to say that he couldn't see an easy path for his two younger sisters.

'There's child marriage in our community,' he said, 'and it's difficult to make people change. It's important for boys to wait until they're twenty-one for marriage and eighteen for girls. I learned this in training, but it's not possible to make everyone understand that child marriage is a bad thing.'

'Have you tried to explain the reasons to wait?' Binod asked him.

'Yes,' Sanoj insisted. 'When my sister was getting married, I tried to tell my grandfather and father to not do it, that she was too young, but there was so much pressure.' Maybe it is the father who exerts pressure, or the village, or maybe, according to Sanoj, 'the girl understands the weight on the father's head in that situation', and does what she can to be a good daughter, quietly accepting the fate of marriage. Girls too carry a burden of responsibility in India. They wear the weight of wedding gold and flowers and expected subservience upon their heads when they are paired with their husbands, even if matched with the sweet-seeming boys that sat before me.

And fathers feel pressure to marry their daughters at a young age to keep their dowry payment from the girl's family to the boy's small. If a girl is older and educated, she requires an older and more educated match, so an older bride translates to a greater dowry. With these escalating dowries, poor parents can't afford to wait. Like child marriage and sex determination tests, dowries are illegal yet routine. Sanoj's sister was married at fifteen. She now lived at her in-laws, her brother said. Other boys countered Sanoj's story with more encouraging anecdotes, but it was the girls who talked most passionately. Several spoke of becoming more comfortable with their bodies after having been taught what was happening to them, after resolving at least some of the mysteries of adolescence. They had lost their shyness by learning about the details of biology.

'They call us dirty because of our period,' said one fourteen-year-old girl named Shruti, the eldest of five siblings. She spoke openly, unaffected by the gaze of the boys in the room, though she did look down at her hands. Her hair was pulled back into a single braid, leaving dark wisps around her forehead. 'It's like we're untouchable for those five days,' she said. 'You feel revolting. You think of yourself as an untouchable. But it is a natural thing.'

Shruti picked at the remnants of red polish on her

fingernails. 'We couldn't talk previously,' she said, suddenly looking up. 'We were very shy. Now that I have training, I can talk and explain this to anyone.' Still, she said, knowledge didn't prevent the snide remarks about menstruation she and other girls faced from their family members and communities, or the forceful jibes about getting married and having a son, or the way women seemed to get blamed for everything. Underlying her words was one question, articulated in a variety of ways by the girls in the room: Why does Indian society have such endless obligations for girls, when boys are free to do what they want?

Another girl took a deep breath before she spoke. She wore a blue salwar and her hair hung down. She leaned forward.

'Today I have a golden chance,' she began, sounding excited to practise her limited English. 'My name is Sandhya Kumari, and I am from Atiyar. I am in class ten and have three sisters and one brother—a simple family.' Slipping back into Hindi, she explained that she and one of her sisters had to sneak away without their mother's permission to attend Pathfinder's training. They had been nervous about getting caught. Sandhya was sixteen, and lamented her life in the village. 'Whenever we go out of the house, the taunting begins,' she said, 'and the gossip about us. My mother said, "After marriage, no one will gossip about you." But is it assured?'

A few girls murmured in assent.

Next to speak were a couple of girls who had walked two hours from their village to reach the class, carrying their school binders and Pathfinder training books across a broad, shallow river and around a large hill. Like most of the other girls in the room they too were Kumaris.

'When I went to training, my mother said it was useless. She told me not to go,' one girl, Asha, said. She had wavy hair and a salwar the colour of a dark eggplant. 'But I said, "Whether it's useless or not, I want to go and see the atmosphere, learn the argument and see what it's about." When I got back, my mother scolded me because I hadn't finished my chores at home.'

Asha's mother was not receptive to her daughter's education. 'This is all for saying, not doing,' Asha recalled her mother saying as she searched for prospective grooms. Asha was in tenth grade then, fifteen years old, and she tried to reason and plead to convince her mother to wait.

'I told my mom, the way you married my sisters, when they were eleven, twelve, fifteen, please don't do the same to me. My brothers-in-law are not very progressive. They don't even have jobs.' She convinced her parents to postpone their search for a groom. 'Mother saw I wanted to do a lot of things. I told her, I might not get a job right away, but I should get an education, then get married. I won't get married before I'm eighteen years old.' She was seventeen and so far her parents had relented.

Also from Kanhaul, Shobha and Suman were sisters in a family with seven children. Both described their village as a confined place.

Suman was slender and stubborn. 'We're determined, no matter what difficulties we'll face, none of us will get married before eighteen or nineteen,' she said. 'And we'll only spend our lives with a good husband, as a couple.' She used the Sanskrit word dampati, meaning husband and wife, but spoke of it as more of an egalitarian pairing. 'I won't settle for any less. What do I need to be happy and keep my children happy?'

Shobha was the most self-possessed. Her forehead was marked with a sparkly bindi that matched an S-shaped pendant hanging from her neck. She sat attentively as the others spoke, only once interjecting, 'We should speak the truth.' But once she had the floor, she commanded the room. All distracted chatter stopped.

'"What sort of place is this?" you'll say if you come to my

village. From Bodhgaya, there's a river and a hill, and behind them is the village, like a cave. People were afraid to go inside. Even my father didn't want to stay in the village. It was claustrophobic.' There had been changes, she said. Some villagers now had phones, and one road was being paved.

She learned about the Pathfinder training course from village elders, who said all girls between fifteen and eighteen should attend. 'But my situation was common in the village,' she said, 'where each household might have six sisters, five sisters. So we made a group with at least one girl from each house. We took the training and then we went back home to teach others.' She had to periodically gulp to catch her breath, as though she had been waiting a very long time to speak, and felt the importance of each word.

Shobha said she was able to get a Pathfinder poster of the life cycle of a human being passing from birth through adolescence, followed by marriage and a young couple weighing birth-control options, and later holding a child as it is being immunized. She used the poster to begin talking with others in the village. Pinki and Binod exchanged looks; they had no idea that one of their students had gone rogue and appointed herself as trainer.

When Shobha finished, Pinki asked if she would continue to work with Pathfinder to organize more training courses. Shobha eagerly agreed.

We emerged from the small schoolhouse into an afternoon damp with mist. Though their hair was neatly pulled back and their clothes were modest, I saw the girls as powerful goddesses, devis eager for justice, who stepped in where the male gods were failing, determined to quietly, or not so quietly, dismantle a world that treats them as second-rate citizens.

Still, the boys and girls clustered in separate packs. Indian boys and girls are intimate within their gendered ranks. Boys intertwine their fingers as they walk down city streets, girls play with each other's bangles and hold hands. But everything in a boy's world suggests that he is favoured even before birth. And what opportunities await twenty-first-century boys in the Buddha's place of enlightenment?

In 2013, journalist Praveen Swami described India's transforming economy as one in which 'son-worshipping parents' have 'coddled ... a mass of young, prospect-less men' into adulthood.[37] Whatever work is available to them pays minimally. Long before India's economy began to slump in 2011, at least one male from every Bihari household sought work in another state, even if he was still a boy. Advertisements and Bollywood films portray India's male elites enjoying flashy motorbikes, glossy-tressed girls and chandelier-lit dining tables. But this lifestyle is as out of reach as it is tantalizing. A quarter of rural households have no computer, mobile phone, landline, transistor radio, bike, motorbike, or car according to the 2011 census.[38] The majority of the population has acquired more material security, but as the income gap widens, resentment grows. Kapila Vatsyayan, who has spent a lifetime studying India's cultural traditions, told author Pavan Varma, 'In all seriousness ... the only enduring emotion (sthayi bhava) among Indians is envy (irsha).'[39]

Boys leave home to labour on a Punjabi farm or at a job site in New Delhi, but too many girls lack even that opportunity in life. For every 1,000 boys born in Bihar, there are only 916 girls. Nationally, 940 girls are born for every 1,000 boys. In Europe and North America, the number is 952.[40] In some Indian states, the sex ratio is even worse. According to the 2011 census, Haryana had only 879 females for every 1,000 males.[41]

These are the 'missing women', as Nobel laureate and economist Amartya Sen described them in 1990, and their absence is a result of what Mara Hvistendahl wrote of in

Unnatural Selection and what inspired the title of Elisabeth Bumiller's book *May You Be the Mother of a Hundred Sons.* Even taking into account the fact that the natural human sex ratio is skewed a bit towards the male, these figures paint a clear portrait of gender-selective-infanticide in India, particularly in the western and northern states. Half of the missing women in India are missing from Bihar and other BIMARU states.[42]

Where have they gone? In the centuries before sex-determination tests, the only option available to parents of unwanted daughters was infanticide: ritually drowning the girl in a tub of milk[43] or feeding her the sticky white sap of the poisonous erukkampal plant, a gangly green bush that grows along Indian roadsides.[44] In Punjab, the death would be accompanied by a chant:

Eat the jaggery
Spin the cotton
You should not come
Send your brother.[45]

But technology changed all that for those who could afford it. Two generations have passed since amniocentesis and chorionic villus sampling first made it possible for a pregnant mother to know the sex of the fetus growing within her. A single generation later, ultrasound became so inexpensive and so portable that all but the poorest expectant parents could make an informed choice about whether to abort. Abortion has been legal since 1971, and in India, even with the rise of its own 'Religious Right', there is no cultural stigma against or political controversy surrounding the procedure. Advertisers once played to the fears of expectant parents about the looming prospect of a dowry. Posters plastered on buses and shop walls in the mid-1980s coaxed women to medical clinics for a sex test with 'Better 500 rupees now than 5,00,000 later'.[46] The sex test, however, was banned in 1994.[47] Still, 'selective abortion of girls, especially for pregnancies after a firstborn girl', according to research published in *The Lancet* in 2011, 'has increased substantially in India. Most of India's population now live in states where selective abortion of girls is common.'[48] The rate is highest in wealthy and educated households. A 2006 study made the conservative estimate that half a million sex-selective abortions occur in India each year.[49] In a New Delhi court in 2014, a judge stated that sex-determination clinics were 'flourishing'.[50]

But even the females who make it past birth face a life of ongoing neglect.[51] A sick boy is twice as likely to be taken to the doctor as is his sister, who is often tended to only at home.[52] Boys are given proper rounds of immunizations and meals even when food is limited, while girls are passed by.[53] One by one, girls perish, while boys grow.

Sunny Hundal, author of *India Dishonoured: Behind a Nation's War on Women,* wrote that 'the number of women being eliminated is on a genocidal scale'. Hundal figured that the total 'missing' is around fifty million over the last three generations, 'thanks to abortions, infanticide, dowry deaths and other kinds of murders'. In the two generations since sex-determination tests became widespread, boys have grown up without their sisters or their future wives. India's skewed gender ratio means that there is a growing class of permanently stranded bachelors, left without any prospect of a spouse. 'By 2020, the country will have around an extra 28 million men of marriageable age,' wrote Hundal. 'For context, that is more than the entire male population of England. Given that India and China together represent around 36% of the entire world's population, the sex-ratio imbalance is unprecedented in human history and likely to have global implications.' In some areas, this lack of women has already led to an increase in sex trafficking. More and more new brides are being sexually 'shared' with the husband's brothers.[54]

As previously mentioned, both men and women tend to

travel in packs in gender-segregated India, but males make riskier decisions because of what is known in social psychology as group polarization: the tendency of individuals in a group to act more extremely than they would on their own. Young Indian bachelors move more freely, and are thus more likely to gather at shops, chai stands and job sites and carouse on nights out. The stage is set for a 'risky shift', when the decision of an otherwise sensible individual is influenced by dangerous group think. The men who hired that New Delhi bus for a night out were just joy riders supposedly looking for some fun, metal bar in hand. Years later, one of the accused was unrepentant. 'A girl is far more responsible for rape than a boy,' he said from jail. 'When being raped, she shouldn't fight back. She should just be silent and allow the rape.'[55]

'For many men,' wrote Praveen Swami, 'violence against women works much as drugs do for addicts: It offers at least the illusion of empowerment where none exists, fixing feelings of rage and impotence.'[56]

THE LINE IS NEVER STRAIGHT

Before and after my trip to Bodhgaya, I sought out every entity working on population and reproductive health in and around Patna. There had been the cheerful air in the Pathfinder office filled with youthful trainers, but also the intensity of a health clinic run by the non-profit Janani, where I watched a line of women lift their legs to a doctor inserting IUDs at a rapid-fire pace. The doctor lingered while working on one woman. She was giving her an abortion. I had seen the squalor of a nearly defunct public health clinic where I waited for an hour to meet a doctor on duty, only to find he hadn't shown up for his shift. He was across the street at his bustling private clinic instead.

And I'd spent countless hours sitting in the nondescript offices of Population Services International, UNFPA and the

National Rural Health Mission (NRHM), talking to numerous bureaucrats and staff of non-profit organizations. One of the Patna interviews had been with Rafay Eajaz Hussain, state programme manager for the Population Foundation of India.[57] He was a towering figure in a khaki, short-sleeved button-down shirt. He offered me mango juice and chips as we sat down to talk.

'Until 2009, we were able to only utilize 28 per cent of the funds we were receiving in the state [of Bihar],' he said. 'By 2011, we reached 85 per cent.' The increase was impressive, but with so much work to do, how was it possible to still operate under-budget? There was abundant federal money available, Hussain insisted—to hire nurses, health workers and teachers—but there weren't enough educated applicants within the state. And few who were qualified from other parts of India were willing to move to lawless and luckless Bihar.

What, I asked, would he choose if he could do only one thing to address population growth?

'Delay the age of marriage, delay the age at first birth, and the spacing between the first and second child,' he began. 'Make quality health services accessible to all through the government system and address the adverse sex ratio. Besides that, education, livelihood, social development, class- and caste-based equity. Bridge the huge gaps in health resources. It will take some time, but one should choose the right path. One needs to improve the system and also address the social and systemic ills like the dowry system and casteism. Women should be respected. Women's empowerment is a major issue. If you give the power to women, they will decide to limit the family, because they have to bear the burden of raising the child.'

I heard myself laugh at this 'one thing', though I wanted to weep. How easy it is to forget that those statistics are made up of individual lives: Asha, Suman, Shobha.

'We might not be able to see the change in our lifetime,'

Rafay continued, 'but we should try to bring the processes on the right path. Someone else will take it forward.'

He shrugged a bit.

'The line,' he said, 'is never straight.'

FROM BEHIND THE BIG DESK

In the Sheikhpura area of Patna, past rows of slatted stalls where rickshaw-wallahs fill their rickshaws with heaps of cauliflowers and brinjals bound for stands around the city, is a complex of government buildings. I arrived there on a Wednesday morning at the same time a rowdy group of broom-wielding protestors were storming the State Institute of Health and Family Welfare, demanding retroviral medicines for people with HIV. A pack of a half-dozen male photojournalists scrambled around the building in an effort to get a good angle, but they lost interest once the shouting crowd crammed itself into a narrow hallway. A security guard watched listlessly until three police officers arrived, one with a rifle slung over his shoulder. I headed over to the adjacent State Health Society building, past a rusting hulk of a bus and a chai stand. I was late.

I was meeting Sanjay Kumar, the executive director of the State Health Society in Bihar, the state's arm of the NRHM.[58] Deferential subordinates slinked in and out of his door until I was led into his palatial office. I was offered a chair in front of his voluminous desk, a dozen chairs set up behind me, as though a classroom of students had just filed out. A tropical fish swam in circles in an aquamarine tank and a Buddha's head stood perched on a stand. A large flat-screen television was playing the news channel Times Now, broadcasting live President Barack Obama's acceptance speech just hours after he had secured office for a second term.

Kumar sat relaxed in his chair, glasses off, behind his citadel of a desk. We figured out that, while I was in high school in

New Jersey, he had been studying at Stonybrook University in nearby New York. But he was a Patna boy, born and bred. And like the others I interviewed, he was full of statistics.

'Our TFR'—total fertility rate—'is not one of the highest in India, it is *the* highest, at 3.7,' he began.

'The good thing is it was stuck at 3.9, and at least we were able to get a .2 TFR reduction, which we hope to do again each year,' he said optimistically. But the government faced multiple challenges in helping its citizens control their reproductive lives. He told me that the nearly 40 per cent of couples who want to use birth control can't access it. 'The government can't keep up with the need, and our primary health care centres are not properly stocked.'

Patna girls, he explained, were getting married at a very early age. Nearly half were already married by the time they turned sixteen. A few years later, most became mothers. 'The single determinant that is affecting the fertility behaviour of women,' he said, 'is education.'

Even with the high fertility rate in Bihar, women with ten years of education gave birth to only two children on average, he said. Those with twelve years of education, both in Bihar and across India, brought the average down further still. Multiple studies have shown that the single most influencing determinant for the number of children a woman will have in her lifetime—superseding race, religion, nationality and class—is her education level.[59] Education as birth control. Education *is* birth control.

Kumar had aspirations for Bihar. NRHM was working on opening high schools closer to villages, so that girls wouldn't drop out because of travel hindrances. In 2006, the Bihar government gave bicycles to girls to help them get to secondary schools, and attendance increased 30 per cent.[60] NRHM was also bringing family counsellors into birthing centres to encourage women who'd just had a child to consider long-term forms of birth control like IUDs that would help them delay their next. Getting women to use reproductive services *before* they'd had three or four children was the only way the TFR would decrease, he said.

We spoke of Pathfinder's PRACHAR programme, which seemed to be highly effective but was ending. PRACHAR had been a three-phase, ten-year project co-funded by the David and Lucille Packard Foundation, the UNFPA,[61] the Government of Bihar and local non-profit partners. By the time Pathfinder was showing me around, the ten-year period was over and funds were drying up. What would happen to the decade of accumulated data, the networks of youth that had been established, all those educational posters?

Kumar didn't dismiss Pathfinder's work, but he stressed its limitations. 'Their programme is only five blocks in the Gaya district. There are 533 blocks in Bihar. If you're going to do a programme in five blocks for ten years, and have no scope for scaling up, it's not going to have the desired effect.'

'Is it the non-profit's job to scale up or the government's?' I asked him. Pathfinder, with a lot of foreign money, had done a great job determining what were the most effective forms of education. When would the government play its part?

'I've told them, you've done this training for so many years, make a template, make a standard operating principle for how this is to be done,' he told me. 'If they've made such a thing, I'm not aware of it.'

Binod Singh had told me that Pathfinder was trying, unsuccessfully it seemed, to get the state to step up. He said Pathfinder had passed along all its teaching materials, yet neither Sanjay Kumar nor others in his department seemed to know anything about this.

For all his resignation about programmes like PRACHAR, Kumar did seem to genuinely care about women's equality.

'RCH'—reproductive and child health—'is nothing but

gender equality,' he said. 'You may put in the best services, but if the gender balance or imbalance remains as it is, perhaps it won't be of much use to women. Women must come out and access these services, and that is going to come only when the gender balance is restored in this society by way of more education, by way of more economic empowerment. And also, I would say, political empowerment. Our problems,' he acknowledged, 'are very acute.'

He recognized the impact of the growing population, the bodies and lives created out of the ether, hungry and dependent on what their parents, their country, earth and water could give them to survive.

Kumar had seen a big change in Bihar since the political shift in 2006, but he wished, he said, that he'd seen it a long while back. He laughed and shrugged. The television showed President Obama reaching the end of his address. *We're not as cynical as the pundits believe,* he said in his distinct cadence. *We are greater than the sum of our individual ambitions…*

'There are advantages of beginning late. You can learn from the best examples everywhere. But there are also disadvantages of beginning late. In spite of all the changes taking place, our baseline still remains low.' Change for the girls of Bihar was heading in the right direction, he believed, but it was not going fast enough.

'Time is the only thing,' he said, 'that is not on our side.'

IT IS IMPOSSIBLE

Back in the one-room schoolhouse, Pinki Kumari and Binod Singh had shown me the shining stars of the PRACHAR programme, but what of the control group, all those girls and boys who lived in villages beyond the modest budget or five-block range of Pathfinder's trainings?

In Jaigir, a village of a few thousand close to Bodhgaya, there had been no three-day training session. Pinki, Binod and

I visited on a day when rains had left the village paths mired in mud. A squadron of young barefoot children wearing a mélange of Western-style clothing followed us like a shadow. A digging tool made of timbers, like something from an earlier century, leaned against a fence woven from sticks. A cart was fashioned out of bamboo and lashed with jute rope. A goat was tethered in a shed full of plumbing paraphernalia. We arrived as outsiders, unrecognized, closely scrutinized from windows and doorways.

Arriving at a house where a group of a dozen girls had gathered to meet us, we rinsed the mud off our feet at a well and entered the room full of girls. Dressed in salwars and saris, they sat on a pale rug spread upon the dirt floor. As visitors, the three of us were directed to plastic chairs, and behind us sat more adults on an elevated bed frame while yet more villagers leaned through doors and windows to observe. The walls were a rough, red brick. Ears of corn hung drying from the ceiling. The girls were slightly younger than the ones we'd met in Bodhgaya, but most of them were already married, as signalled by the touches of vermillion powder along the parts in their hairlines.

Binod introduced us and asked a few preliminary questions to assess the group. The girls were solemn and subdued as they answered his questions. The average age of marriage was fourteen or fifteen. They told us they wanted three children, ideally two sons and a daughter. India's efforts, since the mid-1980s, to promote 'A Small Family Is a Happy Family',[62] has generally worked. It is common to see 'Hum do, humare do'—'We Two, Our Two'—painted on the sides of trucks. But parents still prefer boys, the sooner the better. The girls told us that they'd be scolded if they didn't have a child soon after marriage.

'Have you heard of any options for delaying a child?' Binod asked them. They shook their heads. One older woman sitting on the platform behind us said she'd heard of options from the medical store nearby—a shot and a pill, maybe—but they cost money.

'Are there unmarried girls here who are eager to get married?' Binod asked, his charm falling flat with this crowd. There was a universal, if cautious, 'no' from the group. I glanced at the men standing within earshot, wondering whose fathers were present.

When Binod pressed the girls for more, one spoke up. 'No, we don't want to get married, because we won't be able to continue school.'

'Can you convince your in-laws to let you continue to go?' he asked.

'No, it is impossible,' she said. A goat bleated. A baby cried.

'Can you convince your parents to let you wait for marriage?' he asked.

'No, they want it done,' she said.

Another girl in the group stood out though she didn't speak much. Her mouth was set tightly and her gaze was penetrating. A single small stud dotted her left nostril. Ruby Kumari took everything in. She sat next to her friend Shobha, but they were reserved, showing none of the affection I saw between the Pathfinder-trained girls who played with each other's bangles and leaned on each other as they spoke. Both Ruby and Shobha were fourteen years old. Shobha was married. Ruby was not.

Binod told the girls that Pathfinder could send someone to train them about these matters.

'Can you give them some lunch?' Pinki asked them.

The girls perked up. 'Yes, yes' they said, a little louder, Ruby's voice among them.

'I need a list of names,' Pinki announced, 'and a volunteer.'

Ruby Kumari's slender arm shot into the air.

Most of the girls left and we continued to talk to the women who'd been sitting on the bed platform behind us. One was Gita Kumari, an accredited social health activist (ASHA). ASHAs are the foot soldiers of governmental health efforts, responsible for distributing free or subsidized condoms and birth control pills to people in the villages. Working for Rs 1,200 per month, they encourage vaccinations and get direct remuneration for referrals to health clinics that insert an IUD or perform a sterilization. I met ASHAs who were passionate about their work, who demonstrated how to put a condom on a wooden phallus and took me to meet young couples who spoke unabashedly about their birth control methods, even in front of watchful in-laws. Gita was not one of them.

Wearing a tan dupatta wrapped over her head, and a delicate chain of gold that looped around the tops of her ears from the studs on her lobes, she had been silent when Binod asked the group about available ways to delay pregnancy.

'People don't think about the future,' Gita told us. 'They just want to get rid of the daughters. Sometimes there are four to five in each family, and they want to marry them as soon as possible.' Of the thousand people in the village under her jurisdiction, she said she had successfully enticed only ten or twenty to use condoms, and a half dozen to use IUDs.

Gita and the women around her were all in their late twenties. Most had been married in their mid-teens, had borne two or three children, and then got sterilized.

'It is good to wait for five years after marriage to have a child,' said one woman, 'so the girl's health can improve.'

'No, it is important to have children as soon as possible,' responded Gita, 'otherwise the in-laws will start scolding them.' This was precisely what the girls had said earlier, and precisely the opposite of the message she was supposed to be sending to villagers as their ASHA.

'We wish that they should delay the first child,' said another woman, 'but it is not possible. There is too much pressure from the in-laws. They'll tell the girl's family, we'll give her back and marry our son to someone else.'

'Who is responsible if the child is a girl or boy?' Binod asked them, and one woman said, 'It is God.' But another countered that it was only the male's contribution that was responsible for a son.

'No, it's only God,' the first woman repeated.

A girl standing near us spoke up, 'No, it's up to the man.'

'If it's the man,' Pinki joked, 'then why should the bride not leave the marriage, so she can find a husband who can give her sons?'

No, no, they all said, that is not possible. On that, they were all in agreement.

Gita and the other women stepped out of the house as a few men moved in to replace them on the platform. While the women had talked about social pressures, the men spoke of simple economics. It's best to marry young, they said, when the dowry is still low. To secure a landholder or government official on any level can require a dowry in the range of Rs 6 lakh from the bride's family, they said. The teenage village boys that were standing nearby could be had for a fraction of that, about Rs 25,000.

'How much has the dowry come up in price in the last generation?' I asked the men. They looked at each other and began to laugh. They were in their forties, at least twenty years older than when they were married.

'We had no dowry!' they said.

'What changed?' I asked, puzzled, a victim of the common belief that dowries are an age-old tradition in India.

Poverty and illiteracy were so widespread, one man explained, that a girl's parents would send her off with a couple hundred rupees as a gift when she got married. That was it. But when men began to get more educated, their families started to ask—well, really demand—more. Dowries rose for everyone. Though the practice has been illegal since 1961,[63] it is not only common but it is becoming more widespread and more extravagant, a dark side of India's increasing affluence.

It's business, the men said.

The gift became greed.

LETTER IN A POCKET

Several times in our days together, Pinki nearly swooned as she talked about the changes she had seen in the girls she worked with. 'They have so much energy,' she told me on the ride back from Bodhgaya, her round cheeks balling up above her dimples. 'They have so many ideas! They just need a platform.'

But did the Pathfinder-trained girls represent a substantial shift? As girls in India become educated and learn to speak out for themselves, will they incite a backlash? Power concedes nothing without a demand, as American abolitionist and former slave Frederick Douglass said. The incidents in New Delhi and beyond suggested that sexual violence in India was intensifying, the victims becoming younger, the violence beyond barbaric.

But had the rapes actually increased? Had the media outlets 'rapidly reinvented themselves as rape-reporting journals,' as Amartya Sen has suggested.[64] After reviewing data from the United Nations Office of Drugs and Crime, Sen found that rape estimates in India were far below those in the United States, the United Kingdom and South Africa, even when accounting for under-reporting on a scale of five to ten. Was all the attention that began in late 2012 due to the simple fact that the New Delhi victim was a middle-class student, as opposed to a Dalit villager? Was it that the key fact that moved Indians to take to the streets made the story front-page news in the

New York Times? What will the lives of Asha, Shobha, Suman and the others look like in ten years? Or even just five? Will they have become part of an egalitarian dampati or will they be hidden at home raising their sex-selected sons, the fruits of their hard-won education withering?

Pinki's own story offers some indication of what might happen. When she told girls to find their own voice, she spoke from experience. She herself had no guides, no encouraging voice, no empowering training session.

Pinki told me her own story during the journey back to Patna. Her mother was twelve when she married. Three years later, she had her first daughter, who died shortly thereafter. Six months later, her mother had a boy. Nine months after his birth came Pinki. With each pregnancy, her mother's health deteriorated.

Pinki recalled moments when her mother's pain was so overwhelming that she could neither cook nor look after her children. 'It was tough for her, taking care of the house and young children. I had to go with my granny because it was too much for my mother.'

Pinki was five when she rejoined her family, which relocated to Patna to be closer to schools and work opportunities.

'Every month she got sick, with heavy bleeding, when I was very young,' Pinki said. 'She couldn't work on those days, unable to take care of the house or us. My brother and I learned to take care of ourselves by nine to ten years of age.'

Meanwhile, her mother gave birth to two more girls, to the endless disappointment of her in-laws. Why weren't there any more boys?

'After the fourth child came, and she was also a girl, my father's parents called my mother's parents and said, "Take her. We don't want her, making only girl children,"' Pinki told me. Her mother remained with her husband, son and three daughters regardless. 'But there was no abortion in the 1980s, no testing for sex, and contraception was not popular,' said Pinki.

In 2000, her mother became so sick she was hospitalized, and the doctors advised her to have a hysterectomy, but the family couldn't afford it.

'The doctors said she was sick because of early and continued childbearing,' Pinki said, 'like her uterus was like a cancerous thing.' Her mother's health would continue to falter until the family saved enough money to pay for a hysterectomy in 2004.

Through it all, both of Pinki's parents worked. Her father opened a small electrical shop in Patna, selling switchboards and spools of wire while her mother worked as a nurse when her health allowed it. Though all the kids went to school, Pinki said, 'We all sacrificed our education for our brother.' The girls shared a desk and schoolbooks to keep costs down. Her parents focused on their son. That is how it works: send the boy to college so that he can get a good job and support his parents in the years to come. Girls marry out, becoming part of the groom's family.

Educating a son is a direct investment by the parents, rewarded with a higher dowry received from his bride and the income earnings that will support them through old age. It is the reverse for girls. There are three powerful disincentives for parents to invest in a daughter's education: actual tuition fees, an increased dowry if she's married older, and the fact that her eventual earnings (should she work) would be expected to go to her husband's household, not return to her parents. It's a greater cost than most poor parents can afford.

While the Kumari family struggled, Pinki excelled in school and her brother floundered. Though he was a grade ahead of her, she tutored him. When she graduated from tenth grade, at sixteen years old, her family wanted to find her a husband.

'At first I was so happy,' she recalled. 'I got to wear a sari and jewellery. And lipstick! I imagined I'd get control over my own life, which I imagined meant that I'd get to go to the movies whenever I wanted and no one from my own family would beat me any more.' She laughed at her youthful delusions.

She met with one potential groom. He asked for Rs 20,000 cash plus jewellery, which seemed a reasonable sum to Pinki's family. But her father had one demand: Pinki's education had to continue. All their lives, Pinki said, her father showed her and her sisters great tenderness and compassion. The suitor's family would not agree to the condition.

'They didn't have any educated girls in their family,' Pinki explained, 'and didn't want to have any educated girls in their family. They didn't want an IAS girl'—a working bureaucrat. 'They wanted a girl who would come and take care of the house.'

There were introductions to other prospects, but Pinki desperately wanted the search to end.

'My dad was very strict, but I knew that he loved me. So even though we were living in the same house, seeing each other every day, I wrote a letter to him,' she said. 'I wrote to him saying, "I don't want to do this marriage. I want to continue to study. I don't want to become Manju Devi [another version of her mother]. I want to do something with my life."' She included a vow: 'I promise,' she wrote, 'that I will not do anything against our family's traditions.'

She hid the letter in his coat pocket before he went to his shop, but he saw it before leaving. She watched him read it, as she hid behind a doorway.

'He began crying as he read it. My mom came, and he showed her the letter, and she was angry. She was squalling, hitting me.' Her mother supported her education, but felt immense pressure to marry off the eldest daughter early, especially since there were two more daughters in waiting. Ultimately, the matchmaking was called off. Pinki's paternal grandparents and extended family refused to associate with her. Pinki and her parents used to return regularly to the village, but were now unwelcome. Even a decade later, the situation hadn't changed much.

'I didn't want my family to have to suffer for my education,' Pinki continued. She began tutoring other children to make extra money as she continued school while her brother started drinking and doing drugs.

'My mother was so upset,' she said. 'They'd invested everything in the boy.'

Pinki's earnings covered her sisters' tuition fees, and Poonam Sahi, a kind benefactor who had helped Pinki's mother before, hired Pinki to do mehendi. Sahi paid Pinki a staggering fee, exactly the amount she needed to enter intermediate school. Pinki eventually got a BA in commerce and began supporting her entire family.

Then her little sister fell in love.

'When my younger sister Guriya was seventeen and I was twenty-three, she said she was done with schooling and wanted to get married,' Pinki said. But traditions dictated that Pinki should be married before her younger sister, in the proper descending order.

As Pinki talked, we drove through a village named Dewan, on our way from Bodhgaya back to Patna. It was marked by a break in the fields only a few blocks long where shops closed in on the narrow road, crowded by kids and livestock and bicycles. As we thumped over speed bumps, Binod lowered his window to gently chastise a child for straying into the road. Girls walked home from school in uniform, purple salwars with lavender dupattas slung over their small shoulders, their hair parted and plaited into long black braids. Binod pointed out a Pathfinder mural on a wall depicting a solemn couple, dressed in full wedding regalia, with educational insets of birth control options—the small T-shape of an IUD on an outstretched palm, a package of birth control pills—lined up along the bottom.

Seven miles from Dewan, across a river, was Pinki's grandmother's house, where she lived for a while when she was a girl. There was no electricity, no telephone and no road.

Pinki no longer held fantasies about what marriage meant. 'I'd seen my mother's life, and it was not a fantasy,' she said. Though unhappy, she agreed (in keeping with her promise) to follow tradition and marry before her sister, grateful for the extra years she'd had to pursue her education. She had a degree and was working with a chartered accountant. A few months after her parents began the search, and with Rs 1.5 lakh of her own money, her marriage was fixed.

After marriage, Pinki moved to her in-laws, who 'were still cross that the dowry was too cheap. They said their son was fair and good-looking and I was dark and not pretty enough. But his mother, having no daughters, started to like me. I was able to convince her to let me go back to work.'

It was around this time that Pinki quit her accounting job and started working with Binod at Pathfinder. Her new husband was supportive. He even helped Pinki pursue a master's degree.

She paused. Then she recalled the day that she came home from work and her in-laws told her that her husband was out on a walk. They heard gunfire and a commotion and ran outside into a crowd. Her neighbours urged her away, but she stepped forward. She heard someone calling her husband's name, and when she saw him he was lying on the ground with dozens of bullet wounds perforating his young body. The crowd disappeared, a common Indian response to street crime.

'Everyone turned off their lights and hid,' to avoid getting involved with the police, she said, 'but there was one rickshaw puller, and I begged him to help us get to the hospital. Nobody else would help me.' She took the bleeding form of her husband to the nearby hospital but there was nothing to be done. They'd been married for only a year.

Pinki reached for her water bottle to catch her breath. Binod picked up the story to say what a beautiful new bride Pinki had been, especially at the celebration of Teej, a playful festival where a swing is set up in the courtyard and wives get the day off from chores, dress up in new clothes and pray for their husbands' long lives. She had looked so fine, Binod said, with mehendi designs on her palms and the lovely jewellery around her neck. Pinki stared out of the window, where fields blurred together, occasionally broken by the flash of yellow-flowered flame trees in bloom.

A mere year after she had surrendered, celebrated and stashed the wedding gold, Pinki was a widow in a country that shows little sympathy for them. Though they are no longer completely shunned—they don't have to shave their heads or forgo colourful saris and bangles on their wrists like my great-aunt once did—remarriage is still rarely an option.

Pinki's husband's murder remained unsolved, though Pinki believed it stemmed from a property dispute. She opened an investigation, but her in-laws had no desire to pursue it and her lawyer advised her to drop the case to avoid harassment. Years later, the investigation was still stalled, and nobody expected anything to come of it.

'After the accident my father was paralysed by the news. He felt guilty for having arranged the marriage,' Pinki said. Her brother was an alcoholic, her mother continued struggling with her illness. 'But I was determined.' She paused again, then added, 'Binod pushed me. He was the one who said you must come back from this stigma.'

'She became so dedicated to her work,' Binod said. 'It was—do you know this word?—sublimation. All of her pains—physical, emotional, mental, social—they all went to work.'

Binod nurtured the determined audacity he'd seen in her at that first interview.

'She hasn't told you.' Binod said, 'She had a diploma in classic Katta dance. She had done theatre. And she could shoot a gun!' These things had obviously impressed him. Pathfinder helped her apply for a Ford Foundation fellowship. She got it, and went to the United Kingdom. She earned a master's in public health at the University of Leeds. Her thesis was on youth and sexual reproductive health in Bihar.

The possibility of another marriage remained complicated. There was a proposal from an Englishman, she said, but her priorities were on her schooling. Another courtship with an Indian man back home—partly arranged through friends—brought her back to Bihar, but ultimately left her heartbroken.

'Binod advised me,' Pinki said, '"Why are you upset about this opportunistic man? This foolish man? You have your own life. I thought you were different, not just wanting a husband and children. But this?" They were good words to hear.'

Pinki continued her work with Pathfinder, energized by the adolescents she trained but also unsure of what would become of the girls she helped empower. She wasn't against marriage. She was just trying to teach girls how to delay it. Yet, at home, she said, 'I can't even counsel my own sister.' While Pinki's middle sister sought out marriage before she was eighteen, the youngest was fiercely resistant. Nisha, who was twenty-three when I met Pinki, was more interested in pursuing her career as a folk singer than as a wife.

'When we urge her to go for marriage,' Pinki said, her sister retorts, 'What is good in marriage? See mother's life, or your life. Better to live single and enjoy your life. Everyone will love you if you are able to do something for your family, but then I'll get older and I'll be a useless thing in their house, so why are you forcing me into marriage?'

'The younger generation,' Pinki said, 'is less willing to take up traditional family life. They have not learned positive things from the women's struggle—their mothers' and their sisters'. They have only learned negative things. The male is the person who will rule over you.'

'Do you think men are changing?' I asked her.

She paused for a moment. 'No one knows how people really act when they go home. How they treat their wives.'

Pinki felt women were as much to blame as men. Women raised their sons, instilling in them the expectations of society.

'The mother and the sister are the first who teach the male child how to ruin and rule a woman,' Pinki said. 'They always teach that you are the male, you are first.' She described knowing families with fewer females, and felt that the men within them were more respectful to women. In families with lots of females, she said, 'The grandmother is encouraging him to beat the sister, and the mother tells the father, "Let the boy go to the cinema because there is no one to rape him." I am amazed. These men were like, "You cannot eat before me!" They have been taught, "You are a man. You have the right. No one is to stop you. You are the protector of a girl."' That mindset has proven too easily to veer from protector to controller. The resistance some of the girls got from their mothers about going to the PRACHAR training reflected these same biases, originating not from domineering fathers, but traditional mothers.

Pinki said that through her work with Pathfinder, she has learned that any time women find a place where they can learn, they will speak up. But males love to loiter their days away, stand in front of small shops where they can get a cigarette and just chat and taunt girls. 'They never support any sort of social thing. They are not willing. Whenever you ask a boy what they want to do, he will say he wants to become a doctor, engineer—or some such big dream! You ask a girl, and she will say, "I want to do something with my life. I want to become self-dependent." Whatever you offer to them, they are ready to take it and establish their identity. Girls are doing better everywhere—in school, in family.'

I asked her if she would marry again, and she laughed awkwardly. 'No,' she said, then hedged. 'It's not decided. But if I did, I have one condition. I cannot leave my family. I have to help them financially. A man would not have to help my family, but he could not stop me from helping my family.' She looked out of the window. 'Everyone wants a girl who's working, so she can help him,' contributing domestic labour and wage earnings. 'But I'm not willing to compromise.'

Early the next year, Pinki Kumari made the decision to remarry. It was not an easy one, she told me as she filled me in over a long-distance phone call a couple of years later. 'My husband's family is,' she said 'in one word, a *traditional* family. They are a typical traditional family where you have to cope with so many things.' There were expectations to care for her in-laws and her husband and the new house she found herself living in with them.

Yet she continued to work. She had left Pathfinder and joined Population Foundation of India, continuing her family planning advocacy but now on a policy level. She said she missed the adolescents she used to work with but she was excited about the direction of her work, and still hoped to one day get to the United States and earn a PhD that would allow her to do even more.

As for her earnings, they went to her own family, an act that pleased neither her new husband nor her in-laws, but she said she didn't really care.

'I decided what I am doing because I am strong,' she said, her determination clear even across the crackling line. 'I have the capacity to make my decision. But sometimes I think about those girls who cannot. I think about how painfully they are living.'

Her younger sister never did get married but would consider it, after she gets her master's degree. Her parents were aging but well, and her brother was exactly the same.

I asked her if she was continuing to see change in Bihar and she said yes. 'Even now, if you ask a schoolgirl studying in sixth standard about politics, she'll have an answer,' Pinki said. 'This is a big change. Everyone is talking about politics and development. They are keeping their eyes on these things.'

In the summer of 2014, Pinki and her husband had a baby girl. She plans to have no more children. Her daughter's name is Vedita, she told me, which means 'wisdom' but they call her Vincy. Across the bad connection, it almost sounds like invincible.

TOGETHER

India is undergoing a radical test. Girls from all over South Asia are leaning in, tipping the balance, and hairline fractures are appearing in the ingrained system of chauvinism. Whether sexual violence is on the rise or decline is difficult to know. Whether the aggression is men's bitter reaction to the power they perceive they are losing to women is likewise uncertain.

But what is known is that it is now news. The rise of both women and men who are unwilling to accept the status quo has been startling and encouraging. There are women who bring their daughters into the streets to protest, boys like Sanoj who fight for the rights of their sisters, men like Pinki's father who struggle to educate their daughters.

Everyone is thirsty. Girls and women, after centuries of serving tea to the men in their lives, are reaching for their own cups. I don't want to believe that power is finite. Let the teapot be topped off, let the servings be stretched. Because everyone is striving. In today's India, men and women, boys and girls, share each other's desires for what Pinki calls 'self-independence'. Maybe this is why Pathfinder has found that its trainings are substantially more effective when they teach young men and women simultaneously. It's not just about giving knowledge to the girls or teaching the boys to be respectful. It's about what arises in the ākāśa, that ethereal space between the two sexes. It's about what happens when their lives come together.

The stricter laws against rape that passed at record pace in 2013 might translate into less violence against women. And increasing government support for safety nets and social security could make aging parents less dependent on sons, helping to balance the economic scales that favour a boy child over a girl. Could Shobha or Suman or Pinki's daughters experience in their lifetimes a metamorphosis in India that would allow girls to grow into women whose ambitions, personal and professional, are encouraged, and whose incomes, should they choose to work outside the home, are under their own to control?

Resistance remains. Some local government officials have responded to the rash of rapes by suggesting India lower the legal age of marriage to help curb such crimes. 'Boys and girls should be married by the time they turn sixteen,' they argue, 'so that they do not stray.'[65] In the face of such logic, the problems can seem intractable. But traditions can be lost in just a single generation. So can the beliefs that it is necessary to marry off your daughter at the onset of puberty and that it is her fault if she does not deliver a son, and do so immediately.

I have seen the shift in my own family. My Indian grandmother was married at the age of ten. Her four daughters were married in their late teens and twenties. My father, one of her middle sons, completely broke rank, marrying an American when he was thirty. I wasn't married until the age of forty-four and have chosen not to have children. Among my cousins' grown children who remain in India, arranged marriage remains the norm, but some are holding out against matches they're not willing to accept. Each generation has had fewer children than the one before it, and the levels of education for both males and females tick upward. Our population growth is stable.

But we are a family with relative means. For the vast majority of Indians still struggling to survive, larger structural changes are needed. They are within reach. Kerala once had the highest population growth in India, but since 1971 it has invested heavily in women's education, accessible family planning, and comprehensive health care. With neither threat nor coercion, the fertility rate more than halved in a single generation, from over four to under two.[66] Pinki and Binod reminded me what could be accomplished even in Bihar, a place where the fertility rate for an educated woman was half that of uneducated women.

To stabilize population growth is to rally for literacy, because reading and understanding words on a page develops the same skills needed to read and understand our own bodies. Through this knowledge comes power and autonomy. And speech. The girls I met in Bihar who had their jagriti spoke in feisty voices, their excitement coupled with impatience as they told the stories of their lives. What they found was that learning how to speak—to a husband, a mother-in-law, a doctor, a police officer—is a powerful tool. With this transformation of a private voice into a public voice a public identity is born, one prepared to dissent and stand up for oneself.

'People ask us, "Why do you go to these meetings? Do they give you something?"' Reena Kumari, an eighteen-year-old Bihari girl told me. 'I say, "When you go to pray, do you get something?" They say, "Well, that one girl who did the training met a boy and ran away."' She laughed, and continued speaking quickly, in a strong voice.

'We argue back—you had her for fifteen years and they had

her for three days and you're saying *we* influenced her?' she said. 'There is a flaw in your nurturing, not in our friendship.'

'You fight back with their parents?' I asked.

'*Hum bolti hain!*' she said. 'We speak up! Before training, we didn't know anything, but after, we do. We learned how to find the right words to negotiate. There are so many changes.'

To negotiate such changes is to ask for everything you want, knowing you might only get a fraction. It is to remain unflinching as you look forward into the future of India's women and girls and the generations they will bear. The path ahead is difficult, littered with obstacles, still under construction. But I can imagine the youth I met in Bodhgaya growing up in this new India, their India, moving forward down this road alongside Pinki, with her baby Vincy clutched in her arms. They shape the way as they go. They link their fingers, they quicken their pace, and their voices—rising up into that space between spaces—are unafraid.

CONCLUSION

Faith is the bird that feels the light and sings
when the dawn is still dark.

—Rabindranath Tagore

As I travelled across the subcontinent, I asked nearly everyone I met, 'Do you have hope?' I asked the biologists Munir and Patrick as we sipped tea sitting along the shores of a lake in Rajasthan. I asked activists I met in Bengaluru and epidemiologists from Chennai. I asked Pinki and Binod in the car heading back to Patna, and farmers as we sat in fields in Punjab. A few offered an unequivocal yes, others a resounding no, but most often I received halting replies as people desperately tried to fashion optimistic (but not always convincing) answers.

In Rajasthan, something changed. Kanhaiya and I were standing at the top of the hill behind his house, overlooking the Bhaonta valley, verdant and invigorated by small-scale waterworks. There, under a cerulean sky, I realized I had been asking the wrong question all along.

I was using *hope* as a noun when I should have been using it as a verb, as something active and ongoing. I was using *hope* as a thing to be possessed, something you either had or lacked. Instead of asking, 'Do you have hope?' I should have been asking, 'Do you hope?'

We hope for a bountiful harvest from a Punjabi field that leaves the soil and water intact. He hopes for a vulture's survival as he lets it loose from his hands into the wild. She hopes for a future she begins to imagine on a cloudy afternoon in a one-room schoolhouse in Bihar.

To hope is also to act, and now is India's time for action. Now is the moment to build a new economy that cultivates the country's people and also safeguards its irreplaceable natural resources.

While I worked on this book, India's economy rose and faltered. When the Hindu nationalist Narendra Modi of the Bharatiya Janata Party (BJP) became prime minister in 2014, India was in an economic tailspin. The annual growth rate, racing for the previous two decades, had sputtered. The value of the rupee plummeted while the national deficit soared, along with inflation; food prices escalated out of control, attributed to corruption and mismanagement rather than a lack of food.[1]

Given such dire straits, people fixated on Gujarat, where Modi had been chief minister, as a promised land with a thriving economy, reliable electricity and steady gas supplies.[2] Regardless of the reality of these claims, it was a vision many wanted to see realized across India. Dubbed 'Development Man', Modi has promised Indians everything from bullet trains to well-planned cities. Many citizens await the economic miracle to unfold, desperate for growth and development.

But what direction will India's progress take? 'If we talk of promoting development, what have we in mind, goods or people?' asked E.F. Schumacher in his classic book *Small Is Beautiful,* published during the 1973 energy crisis. 'If it is people—which particular people?'[3]

In economic development, the upper echelon of elites recognizes an opportunity to amass more wealth, and the slim-yet-growing middle class (now just 10 per cent of the population) envisions a continued trajectory of upward mobility. But for the masses who struggle daily for survival, their wants are more elemental. Mostly left behind in the recent growth spurt, in which only a fraction received the mist of trickle-down benefits, these citizens still lack clean water, enough food, or a light to switch on when the sun goes down. India needs development for the 160 million women who cook over lethal wood-fuelled stoves that blacken their lungs and skies. Development must reach the 60 per cent of urban and rural citizens who live beyond the reach of clean water. It must fill the plates of schoolchildren who need more sustenance than what a single free midday meal can provide, and feed the pregnant mothers who don't even receive that small, steady offering.

I often travelled between the places I visited by train. I would leave my seat and lean out of the open doors, watching the landscape slip by as the wind tugged at my hair. From that slightly elevated vantage, I could see the paths that people take. They are desire lines, the worn evidence of the choices we make every day. Byways braided across gullies strewn with refuse, footpaths etched through fields of golden mustard, passages disappearing into dry deciduous forests. Desire lines mark not where people should go, but where they choose to go. It is along desire lines that love or loss or hunger or need call them. Desire lines lead to junctures and, each time, there is a decision to be made. Where to go next?

At this moment, India must decide what her future will be. Two paths present themselves as her development efforts advance. Macro is the way of the past, and of the West. In twentieth-century Europe and North America, to develop was to go big. Rapid advancements in technology enabled humans to build ever more complicated systems to create energy, siphon water and grow food. Mega dams grew to such epic proportions that they drew tourists. Big concrete. Big steel. We split atoms to release previously unimaginable amounts of energy.

Industrialization, not limited to manufacturing, became the model by which to raise meat and grow crops. These efforts were incredibly productive. They also came at untold collateral cost to the environment and human communities, and they have proven unsustainable over the long term.

India could follow the West down this well-worn path of the big, wrenching every last resource from the ecosystems of South Asia and giving back nothing. It takes little imagination to envision such a future: overhead will be the blackened skies of England in the mid-nineteenth century, or America soon after, or Beijing yesterday, or New Delhi today. This addiction to the big will result in more dead zones where the rivers meet the seas, like the ones in the Gulf of Mexico and the Baltic Sea caused from agricultural run-off and pollution. Indian rivers will run so thick with effluent that they burst into flames as Cleveland's Cuyahoga River repeatedly did.

Or India could choose the other path: just as she created her own form of democracy, intrinsically Indian and not just a replica of the Western model, she could create her own way to develop her people and resources.[4] Growth could be guided by environmental self-rule. Call it Eco Swaraj, something only a few developed countries have tried on any large scale and that is nearly non-existent in the developing world. Imagine an India in which innovation comes from the micro, where communities at the grass roots decide what they need and how to get it. A bottom-up approach to India's environmental crises can feed people without depleting the soil, draw water from wells without diminishing aquifers or displacing communities, fuel India's growing energy needs without wrecking the environment. The technology exists, and is continually improving, and, coupled with the traditional knowledge that remains, could be transformative.

I witnessed the small, the local, the indigenous at work across India. Vinod Jyani revitalizes the soil of his seventh-generation farm. Villagers of Bhaonta collect rainwater in johads of their own making. Conservationists in West Bengal save vultures across a subcontinent. A girl educates a village, armed with a poster and new knowledge. 'We have to support our small heroes,' wrote Arundhati Roy. 'Of these we have many.' She too advocated for a Century of the Small, the 'dismantling of the Big. Big bombs, big dams, big ideologies, big contradictions, big countries, big wars, big heroes, big mistakes.'[5] Even 'Development Man' Modi, one year after his election, was holding up small-scale traders, not 'big industrialists', as the force that would drive job growth in India.[6]

Remedies for India's environmental and economic ills will not be found solely in a shovelful of dirt, or an induction stove powered by a rooftop solar panel. The catastrophe of lost biodiversity will not vanish even if the vulture is returned. But today, more than ever, we need to draw more out of less. Small systems require less capital and can be implemented quickly. They allow for adaptation to the rapidly changing needs of individuals and societies in a way that macro systems, expensive and years in the making, simply cannot. To go small is to be agile, the darting swallow instead of the lumbering goose.

We live in an era of mobile apps and start-up companies both high-tech and low-tech, a time of micro-lending and micro-enterprises. The changes are revolutionizing our daily lives. It is a time when the collective power of many small efforts can add up to a cohesive and impactful whole. In an interconnected world, knowledge can reach *out* from remote corners of the world as easily as it can reach into them. With this knowledge comes awareness—and consciousness. My young cousins refuse to set off Diwali fireworks because they know that they cause pollution and are made with child labour. So many of the Indians I met, from all economic classes, care deeply about the impacts of their actions. They understand that there are alternatives to importing the top-down model from the West, and they are ready to implement them.

If successful, this model of the micro could do more than just solve India's woes, it could provide an example for developing behemoths like China, small struggling countries in Africa, and places such as the United States, where the once impressive macro infrastructure of the last century is in crumbling disrepair. In the US, the population pressures are lower and development is more advanced; so the need to resolve environmental problems might seem less urgent. But all the challenges that are visiting India today are around the corner in the Western hemisphere. India has an opportunity not only to construct a path forward of its own making but also to create one that the US, China and other countries can follow to effectively address the compromises of the Green Revolution, the disruptions of mega dams, the thirst of landscapes in drought.

Modi may have to think small to develop India while preserving its environment, but he also needs the support of the Indians who voted for him and those who did not. No matter how clever his nickname, no leader can accomplish this task alone. Indians from Punjab to Tamil Nadu, Karnataka to West Bengal, must shake off the last vestiges of passive colonialism and take control of their own lives and the destiny of their country. 'Think what we can do,' Vinod Jyani said as we crossed his fields in Punjab. 'We can do this,' said Kanhaiya as we sat in his home in Rajasthan. Hai na?

In the summer of 2012, some of India's richest entrepreneurs gathered with Bill Gates to discuss philanthropy.[7] Indian businesses, under increasing pressure to demonstrate some token of responsibility to the communities that support them, are launching corporate social responsibility (CSR) campaigns, and government mandates are broadening their reach.[8] There are indications that some CSR programmes are enacting fundamental change, geared towards making wide and lasting impacts as opposed to, say, just giving out freebie cookstoves that don't work. Another hopeful sign is that the number of philanthropic donors under the age of thirty is increasing, and more than half of the population is under thirty.[9] Young people know that a well-functioning nation must support not only a strong economy but also livable communities.

The macro still has its place. National and international agreements can serve India and our globalized world on the policy level, bringing much-needed uniformity to everything from how best to restrain runaway carbon emissions to codifying what 'organic' means. The consistency and interconnectedness of the big are also an efficient way to disseminate new information, by spreading strategies on which sustainable agricultural techniques garner the highest yields or how to activate the most powerful renewable energy systems. Think neither Jawaharlal Nehru's monolithic top-down industrialization nor Gandhi's austere romantic agrarian ideal. Instead, draw the best from both.

India, home of the god of small things, is the perfect place for the model of the micro to emerge. The concept of small scale is rooted in the South Asian landscape and ethos. India's vast agricultural lands are largely composed of millions of tiny 3-acre farms. Petite yet packed family-owned shops line roads in every town and city. India has not razed entire countrysides (and the natural and human histories they harbour) in order to erect new cities as China has. The arrival of Wal-Mart was a long, slow process and the big box stores still seem like strange aberrations, more marvel than actual market. Instead of trampling the small in favour of the big, development strategies could tap into the micro, cultivate it, embrace its local knowledge and community power. India has tremendous resources, human and natural, and the fact that the country's infrastructure is still so underdeveloped makes it a place ripe with opportunity for such a paradigm shift.

India can build a booming economy from the ground up on a foundation of conservation and efficiency. She could be a pioneer, rewriting the human development script, which has so far followed a rigid hierarchy of human needs that assumes that the poor cannot care for their environment until their bellies are full. But the stories say otherwise. India can produce quintals of wheat and gigawatts of energy while caring for the natural resources that sustain life. For many Indians, rich and poor alike, this is the path they desire.

India is young. India is ready. India is the place for this model of the micro to emerge. India's citizens are known for their jugaad innovation, their creative adaptability, and their capacity to fix what is broken with whatever might happen to be at hand. This is the time to seize the jugaad spirit and use it as a tool rather than a reaction, to produce something more proactive and permanent. In 1947, India achieved political Swaraj after decades of non-violent struggle. Economic reforms initiated in the early 1990s brought a modicum of economic Swaraj that is still emerging. Now is the time for the rest of India to join Vinod Jyani, Kanhaiya and Pinki as they fight just as hard for self-rule of a new kind. Agro Swaraj. Water Swaraj. Energy Swaraj. Sexual Swaraj. Eco Swaraj.

We all carry weighty turbans woven from threads of responsibility upon our heads. But we are not Atlas, lifting, alone, back bent. There are neighbours to steady us and to help us stand. There are women to sing with us. *Hum bolti hain.* 'We speak up,' a group of girls exclaim. A man plants a seed. A woman lifts a shovel. A girl raises her hand.

ACKNOWLEDGEMENTS

The seed of an idea can only grow given the right conditions, and I am forever indebted to all the people who made *Elemental India* possible.

First and foremost, to my parents, Mani and Ruth Subramanian: Thank you for letting me set up a desk in the corner of the family room when I declared myself a writer after reading E.B. White and Roald Dahl when I was still in my single digits, for the open dictionary prominently placed in front of the window, and for supporting all those trips to the libraries along the shore and in multiple universities, from New Orleans to New York. You tricked me into believing I was capable of anything. Thanks especially to my father, who could have left India and never returned, but chose not to and in the process gave me another home.

To the P.R. Mahadevan clan, in India and around the world, who always welcomed me like a native daughter. To the generations before me—especially my beloved late grandparents Thatha and Patti—and to the younger generations who are rising up and shaping the direction of their country. It was all joy to work with my cousin-once-removed, Pavithra Mohanraj: companion, translator, guide, friend and provider of hope for the future of India.

I am lucky to have never not known my brother Ravi Subramanian, dear friend, excellent travel partner, and arch story-collector rival. Harish Ramachandran, philosopher-cousin-brother, you always keep me up past my bedtime but make it worth it.

To all of the Prothero entourage, especially Helen and Dick, and Molly and Lucy, who forgave my many absences from family gatherings as I travelled, and always welcomed me home to Cape Cod, and into a family that knows no bounds of love.

Apologies to all the friends and family who accepted my 'nothing but book' excuses. Let's make up for lost time, shall we?

For Pollyanna Lind, who is always there. Distance means nothing.

It was a pleasure to work with the editors who published earlier versions of some of the stories from these pages. To the excellent Ted Genoways of the *Virginia Quarterly Review (VQR)*, who early on supported my obsession to understand what a collapse of the vulture population might mean to India. The piece he published was instrumental in transforming an amorphous idea about a book on India's environment into *Elemental India.* My thanks also to a trio of brilliant editors: Paul Reyes of *VQR* ('India's Golden Chance'), Rich Monastersky of *Nature* ('Deadly Dinners'), and Basharat Peer of the *New York Times* India Ink ('Bihar School Deaths Highlight India's Struggle with Pesticides'). Thanks also to the editors of *Orion*, *Wall Street Journal*, *Revealer*, *Killing the Buddha*, and now-defunct *Search*, who further helped build bridges between these Indian stories and American readers.

To my literary agents: Russ Galen, who encouraged me early on; Kathleen Anderson, who stepped in at a critical point; and the extraordinarily dedicated Elise Capron of the Sandra Dijkstra Agency, who has cheered me over the line into publication.

A huge hurrah to HarperCollins India editors Karthika V.K. and Ajitha G.S. for your early support and unflagging enthusiasm. You first made me feel like a book author. I am grateful.

At PublicAffairs, my deepest appreciation for Clive Priddle and Maria Goldverg, who revived my faith in the mythical writer–editor relationship I'd heard no longer existed in New York publishing, as well as Michelle Welsh-Horst and the rest of the talented PublicAffairs team, from copy editors to designers. I could barely keep up with you.

Cross-pollination came from more people than I can name. (Please forgive the absences.) My deepest gratitude to Basharat Peer, who prodded me to write about people as well as birds, and Ananya Vajpeyi, writers both. You have inspired me from the steps of Brooklyn brownstones to the turquoise-trimmed haven in Nizamuddin. Thanks also to Rhitu Chatterjee, Mehboob Jelani, Praveen Donthi, Hartosh Singh Bal, and the many other Indian friends and friends-of-friends who were always willing to field my last minute cries for help from far-flung corners of India.

In Oregon, the community of the Aprovecho Research Center and all those who choose to live simply in the land of plenty remain a model of hope and possibility. It was there that I learned how to get my hands dirty, build johads to capture water, and use wood cookstoves and solar cookers to metamorphose Earth's elements into a meal. Also, how to make chocolate and butcher an emu and dance on the ruins with the rest of the riff-raff. It is in Cottage Grove where the elements came to life for me.

At New York University's journalism school, I gleaned how to listen from the late Ellen Willis, how to find my way into the mysterious publishing world from Rob Boynton, and how to unveil the raw soul of a story from Jeff Sharlet, who has served as tough-love guru with just the right amount of grit. It was Jeff who led me into the cabal of the literary magazine *Killing the Buddha*. There, my thanks to Ashley Makar, Nathan Schneider,

Peter Manseau, Quince Mountain, Paul Morris, Brook Wilensky-Lanford, and the other Buddha-killers. Also in New York, thanks to John Hedigan and Amanda Schuster, for books and booze; Matthew Fishbane for being there, (or if not, giving me a key); Ann Neumann, Kathryn Joyce, and the rest of the Brooklyn writers group that let me loiter long after I skipped town. To Tom Scanlon and Laura Harrison, for valuable editorial feedback. My love and gratitude to those with whom I shared a home on Hawthorne Street, where I could always turn to (the late and deeply missed) Matt Power for unfettered enthusiasm, Jessica Benko for a good science geek out, Michael Vazquez for strong coffee and mind-bending conversation, and Par Parekh for joyful play. And to the steady stream of visitors, writers, rabble-rousers, and other residents who passed through, providing fuel for the journalistic fire.

Appreciation also for unknown committee members who selected me for the awards and fellowships that encouraged this book and pushed me to take it further. The United States-India Educational Foundation provided a generous Fulbright-Nehru Senior Research fellowship that gave me the resources for reporting otherwise beyond my freelancer's reach and connected me to Dr Kamna Sachdeva of TERI University. The Society of Environmental Journalists (SEJ) has also been an invaluable resource for me as an independent journalist, its small army of freelancers serving as my water-cooler cyber colleagues and cheerleaders. Thanks also to SEJ's Fund for Environmental Journalism for providing funds for a highly productive reporting trip in 2012.

The Blue Mountain Center, the Metcalf Institute for Marine and Environmental Studies, and Institutes for Journalism and Natural Resources introduced me to all sorts of fellow-travellers in science and art of all forms, including George Brant, Maleea Acker, Mohan Sikka and many others.

Thank you to all who helped me negotiate the multilingual

land that is India, including Pavitra Mohanraj, Shilpi Suneja, Shivani Karhadkar, Remet Khan, the random people I accosted on the street and so many others. Thanks also to the many scientists who helped me translate their data into stories, especially Munir Virani.

India is a complex country, and I will never know it as a native. So I am as grateful as I am indebted to the hundreds of Indians I met on my travels who led me along their desire lines, sharing with me the stories of lives, even when difficult. Thank you, dhanyavad, shukria.

Finally, and always, my thanks to Stephen Prothero, grounding element in my life of eternal movement, endlessly patient and insightful editor, great thinker. Thank you for giving me a reason to say yes.

NOTES

INTRODUCTION

1. Jonathan Ablett, Aadarsh Baijal, Eric Beinhocker, Anupam Bose, Diana Farrell, Ulrich Gersch, Ezra Greenberg, Shishir Gupta and Sumit Gupta, "The 'Bird of Gold": the Rise of India's Consumer Market', McKinsey Global Institute, May 2007, http://www.mckinsey.com/insights/asia-pacific/the_bird_of_gold
2. 'Ernst & Young Rapid Growth Markets Forecast (RGMF)', 24 October 2011, http://www.ey.com/IN/en/Newsroom/News-releases/Ernst-and-Young-Rapid-Growth-Markets-Forecast
3. Patrick French, *India: A Portrait* (New York: Vintage Departures, 2011), 192.
4. 'Census of India 2011: Houselisting and Housing Census Data Highlights—2011', http://www.censusindia.gov.in/2011census/hlo/hlo_highlights.html
5. Dr C. Chandramouli, 'Housing, Household Amenities and Assets—Key Results from Census 2011', Census of India, Government of India, 2012, slide 57 and 58, http://www.censusindia.gov.in/2011census/hlo/Data_sheet/India/00_2011_Housing_India.ppt. 160 million figure from: Gautam N. Yadama, *Fires, Fuel & the Fate of 3 Billion: The State of the Energy Impoverished* (New York: Oxford University Press, 2013), 40, with several references: C. Venkataraman, A.D. Sagar, G. Habib, N. Lam, and K.R. Smith, 'The Indian National Initiative for Advanced Biomass Cookstoves: The Benefits of Clean Combustion', *Energy for Sustainable Development* 14, No. 2 (2010), 63–72; Stephen S. Lim, et al., 'A Comparative Risk Assessment

of Burden of Disease and Injury Attributable to 67 Risk Factors and Risk Factor Clusters in 21 Regions, 1990–2010: A Systematic Analysis for the Global Burden of Disease Study 2010', *The Lancet* 380, (2012), 2224–60; T. Vos et al., 'Years Lived with Disability (YLDs) for 1160 sequelae of 289 diseases and injuries 1990–2010: A Systematic Analysis for the Global Burden of Disease Study 2010', *The Lancet* 380 (2012), 2163–296; and H. Wang et al., 'Age-specific and sex-specific mortality in 187 countries 1970–2010: A Systematic Analysis for the Global Burden of Disease Study 2010', *The Lancet* 380 (2012), 2071–2094.

6. Gardiner Harris, 'Rival Economists in Public Battle over Cure for India's Poverty', *New York Times,* 21 August 2013, http://www.nytimes.com/2013/08/22/world/asia/rival-economists-in-public-battle-over-cure-for-indias-poverty.html?pagewanted=all&_r=0.
7. Sachiko Murata, *The Tao of Islam: A Sourcebook on Gender Relationships in Islamic Thought* (Albany, NY: SUNY Press, 1992), 135–37; and P. Bearman, T. Bianquis, C.E. Bosworth, E. van Donzel, and W.P. Heinrichs, (eds), 'Unsur,' *Encyclopaedia of Islam,* 2nd ed. (Brill Online, 2015), http://referenceworks.brillonline.com/browse/encyclopaedia-of-islam-2/alphaRange/Uh%20-%20Un/U
8. Tenzin Wangyal Rinpoche, *Healing with Form, Energy, and Light* (Ithaca, NY: Snow Lion Publications, 2002), 1.
9. Ronnie Littlejohn, 'Wuxing (Wu-hsing)', Internet Encyclopedia of Philosophy, accessed 4 March 2015, http://www.iep.utm.edu/wuxing/.
10. K.L. Seshagiri Rao, 'The Five Great Elements: An Ecological Perspective', in Christopher Key Chapple and Mary Evelyn Tucker (eds), *Hinduism and Ecology: The Intersection of Earth, Sky and Water* (New Delhi: Oxford University Press, 2000), 26.
11. http://www.npr.org/templates/story/story.php?storyId=122294630
12. http://www.telegraph.co.uk/news/worldnews/africaandindianocean/10305000/Africas-population-to-double-to-2.4-billion-by-2050.html

1. PRITHVI—EARTH

1. 'Nobel Lecture: The Green Revolution, Peace, and Humanity', Nobelprize.org, 11 December 1970, http://www.nobelprize.org/nobel_prizes/peace/laureates/1970/borlaug-lecture.html, accessed on 6 March 2015.
2. Swaran Singh, interview with the author, 4 October 2012.
3. Tom Philpott, 'A Brief History of Our Deadly Addiction to Nitrogen Fertilizer', *Mother Jones,* 19 April 2013, http://www.motherjones.com/tom-philpott/2013/04/history-nitrogen-fertilizer-ammonium-nitrate.
4. Dan Charles, 'Fertilized World', *National Geographic,* May 2013, http://ngm.nationalgeographic.com/2013/05/fertilized-world/charles-text.
5. Kenneth R. Weiss, 'In India, Agriculture's Green Revolution Dries Up', *Los Angeles Times,* 16 November 2009, http://www.latimes.com/world/population/la-fg-population-matters-india-water-20120726-m-html-htmlstory.html.
6. Ibid.
7. Columbia Water Center, 'Punjab, India', Earth Institute, Columbia University, http://water.columbia.edu/research-projects/india/punjab-india/; and 'Wheat—World Supply and Demand Summary,' Spectrum Commodities, accessed 6 March 2015, http://www.spectrumcommodities.com/education/commodity/statistics/wheat.html, accessed on 6 March 2015.
8. Manu Moudgil, 'Punjab Wades in Shallow Waters', India Water Portal, May 3, 2014, http://www.indiawaterportal.org/articles/punjab-wades-low-waters, accessed on 6 March 2015.
9. Anjali Singh and Mandeep Inder Kaur, 'A Health Surveillance of Pesticide Sprayers in Talwandi Sabo Area of Punjab, North-West India', *Journal of Human Ecology* 37, No. 2 (2012), 134.
10. N.S. Tiwana, Neelima Jerath, S.S. Ladhar, Gurharminder Singh, Rajesh Paul, D.K. Dua and H.K. Parwana, 'State of Environment—Punjab 2007', Chandigarh: Punjab State Council for Science and Technology, 2007, 40–41.
11. Brian Murton, 'Famine', in *The Cambridge World History of Food,* Vol. 1 (New York: Cambridge University Press, 2000), 1412.
12. M.S. Swaminathan, *From Green to Evergreen Revolution: Indian*

Agriculture: Performance and Challenges (New Delhi: Academic Foundation, 2010), 17.

13. N.P. Nawani, 'Historical Perspective of Food Management in India', in *Indian Experience on Household Food and Nutrition Security,* a Regional Expert Consultation, FAO Bangkok, 8–11 August 1994, http://www.fao.org/docrep/x0172e/x0172e03.htm, accessed 6 March 2015.
14. Edward Luce, *In Spite of the Gods: The Strange Rise of Modern India* (New York: Anchor Books, 2006), 30.
15. Ibid.
16. Cynthia Graber, 'The Next Green Revolution May Rely on Microbes', *Nova Next, PBS,* 12 June 2014, http://www.pbs.org/wgbh/nova/next/nature/more-food-with-microbes/
17. Ibid.
18. Jennifer Burney and V. Ramanathan, 'Recent Climate and Air Pollution Impacts on Indian Agriculture', *Proceedings of the National Academy of Sciences* 111, No. 46 (2014), 16322, http://www.pnas.org/content/early/2014/10/29/1317275111.
19. Graber, 'The Next Green Revolution.'
20. Ministry of Human Resource Development, 'About the Mid Day Meal Scheme', Mid Day Meal Scheme, Department of School Education and Literacy, Government of India, http://mdm.nic.in/aboutus.html, accessed 6 March 2015.
21. 'Forensic Report Confirms Poisonous Pesticide in Bihar Midday Meal', *The Times of India,* 20 July 2013, http://timesofindia.indiatimes.com/india/Forensic-report-confirms-poisonous-pesticide-in-Bihar-midday-meal/articleshow/21193717.cms
22. P.K. Shetty, 'Socio-Ecological Implications of Pesticide Use in India', *Economic and Political Weekly* 39, No. 49 (December 2004), 5263.
23. '8 Reasons Why Mid-Day Meal Poisoning Pesticides Should Be Banned', *The Economic Times,* http://economictimes.indiatimes.com/slideshows/nation-world/8-reasons-why-mid-day-meal-poisoning-pesticide-should-be-banned/2–76000-people-die-per-year-in-india/slideshow/21423546.cms, accessed on 6 March 2015.
24. Praveen Donthi, 'Cancer Express', *Hindustan Times,* 17 January

2010, http://www.hindustantimes.com/india-news/cancer-express/article1-498286.aspx.

25. David Gunnell and Michael Eddleston, 'Suicide by Intentional Ingestion of Pesticides: A Continuing Tragedy in Developing Countries', *International Journal of Epidemiology* 32, No. 6 (2003), 902, http://ije.oxfordjournals.org/content/32/6/902.full.
26. 'India's Farmer Suicides: Are Deaths Linked to GM Cotton?—In Pictures', *The Guardian,* 4 May 2014, http://www.theguardian.com/global-development/gallery/2014/may/05/india-cotton-suicides-farmer-deaths-gm-seeds, accessed on 6 March 2005.
27. Kate Kelland, 'WHO Had Asked India to Ban Toxin That Killed Children', Reuters, 22 July 2013, http://www.reuters.com/article/2013/07/22/india-poisoning-monocrotophos-idUSL6N0FS13020130722, accessed on 6 March 2005.
28. Sujay Mehdudia, 'Banned Pesticides Being Used in India, Admits Pawar', *The Hindu,* 12 March 2011, http://www.thehindu.com/news/national/banned-pesticides-being-used-in-india-admits-pawar/article1529735.ece.
29. Donthi, 'Cancer Express.'
30. Rashmi Sanghi, M.K.K. Pillai, T.R. Jayalekshmi and A. Nair, 'Organochlorine and Organophosphorus Pesticide Residues in Breast Milk from Bhopal, Madhya Pradesh, India, *Human & Experimental Toxicology* 22, No. 2 (2003), 73, http://connection.ebscohost.com/c/articles/9283970/organochlorine-organophosphorus-pesticide-residues-breast-milk-from-bhopal-madhya-pradesh-india.
31. National Institute of Environmental Health Sciences, 'Endocrine Disruptors', National Institute of Health, US Department of Health and Human Services, http://www.niehs.nih.gov/health/topics/agents/endocrine/, accessed on 6 March 2015.
32. Florence Williams, *Breasts: A Natural and Unnatural History* (New York: W.W. Norton, 2012), 88–93.
33. Neha Madaan, 'Traces of Pesticide in Fruits, Veggies,' *Times of India,* 5 March 2014, http://timesofindia.indiatimes.com/home/science/Traces-of-pesticide-in-fruits-veggies/articleshow/31445588.cms.
34. R. Ramabhadran Pillai, 'High Pesticide Content Found in

Vegetables across Kerala', *The Hindu*, 12 January 2014, http://www.thehindu.com/news/national/kerala/article5569659.ece

35. Ibid.
36. 'Adulteration All Pervasive', *The Daily Star*, 27 May 2014, http://www.thedailystar.net/adulteration-all-pervasive-25830
37. Mark W. Rosegrant, Jawoo Koo, Nicola Cenacchi, Claudia Ringler, Richard Robertson, Myles Fisher, Cindy Cox, Karen Garrett, Nicostrato D. Perez, Pascale Sabbagh, 'Food Security in a World of Natural Resource Scarcity: The Role of Agricultural Technologies', (Washington DC, International Food Policy Research Institute, 2014), 26, citing E. Oerke, 'Crop Losses to Pests', *Journal of Agricultural Science* 144 (2006), 31–43.
38. Paul Wallich, 'Everything You Always Wanted to Know About Big Ag', *IEEE Spectrum*, June 2013, 36.
39. Ibid.
40. Ibid., 34.
41. Weiss, 'In India, Agriculture's Green Revolution.'
42. A. Vaidyanthan, *Water Resources of India* (New Delhi: Oxford University Press, 2013), xxxiii.
43. Moudgil, 'Punjab Wades.'
44. Ibid.
45. Priya Yadav, 'Ground Water Level Declining, Contamination Rising in Punjab', *Times of India*, 29 June 2012, http://timesofindia.indiatimes.com/india/Ground-water-level-declining-contamination-rising-in-Punjab/articleshow/14482758.cms
46. Seema Singh, 'Pumping Punjab Dry', *IEEE Spectrum*, 28 May 2010, http://spectrum.ieee.org/energy/environment/pumping-punjab-dry
47. Moudgil, 'Punjab Wades.'
48. Ibid.
49. Priya Yadav, 'Ground Water Level Declining, Contamination Rising in Punjab.'
50. Ibid.
51. US Environmental Protection Agency (EPA), 'Agricultural Nonpoint Source Fact Sheet', USEPA, http://water.epa.gov/polwaste/nps/agriculture_facts.cfm, accessed on 6 March 2015.

52. Yadav, 'Ground Water Level Declining.'
53. 'Punjab in Grip of Cancer, Over 33,000 Died in Last Five Years: Minister', *The India Express,* 29 January 2013, http://www.indianexpress.com/news/punjab-in-grip-of-cancer-over-33000-died-in-last-five-years-minister/1066072/
54. Rohan Dua, 'No Poll Stop for Cancer Train', *Times of India,* 28 April 2014, http://timesofindia.indiatimes.com/city/chandigarh/No-poll-stop-for-cancer-train/articleshow/34170980.cms
55. Sweta Dutta, 'Bathinda to Bikaner, Aboard "Cancer Train"', *The Indian Express,* 1 December 2013, http://archive.indianexpress.com/news/bathinda-to-bikaner-aboard—cancer-train-/1201723/0
56. Dua, 'No Poll Stop.'
57. Dutta, 'Bathinda to Bikaner, Aboard "Cancer Train"'.
58. Ibid.
59. Umendra Dutt, multiple interviews with the author, 28 September 28–6 October 2012.
60. Jason Bittel, 'Second Silent Spring? Bird Declines Linked to Popular Pesticides', *National Geographic,* 9 July 2014, http://news.nationalgeographic.com/news/2014/07/140709-birds-insects-pesticides-insecticides-neonicotinoids-silent-spring/
61. Sukhdev 'Gora' Singh, interview with author, 28 September 2012.
62. TheCompostGardener.com, 'Actinomycetes,' TheCompost Gardener.com, http://www.the-compost-gardener.com/actinomycetes.html, accessed on 6 March 2015.
63. Amarjeet Sharma, interview with the author, 5 October 2012.
64. Anna Marie Nicolayson, *The Dark Shadow of the Green Revolution: Empowering Small Farmers in India Through Organic Agriculture and Biodiversity Conservation* (Saarbrücken, Germany: Lambert Academic Publishing, 2013), 75, citing *Withering Punjab Agriculture: Can It Regain Its Leadership?* (New Delhi: International Food Policy Research Institute, March 2007).
65. Gordon Conway, *One Billion Hungry: Can We Feed the World?* (Ithaca, NY: Cornell University Press, 2012), 89.
66. Guillaume Gruere, *Bt Cotton and Farmer Suicides in India: Reviewing*

the Evidence, (New Delhi: International Food Policy Research Institute, 2008), 7.

67. Ibid.
68. Susan S. Lang, 'Seven-Year Glitch: Cornell Warns that Chinese GM Cotton Farmers Are Losing Money Due to "Secondary" Pests', *Cornell Chronicle,* Cornell University, 25 July 2006, http://www.news.cornell.edu/stories/2006/07/bt-cotton-china-fails-reap-profit-after-seven-years
69. Lakshman Krishnamurthi and Sugandha Khandelwal, 'Agricultural Journal: China Versus India by the Numbers', *Wall Street Journal,* 20 September 2011, http://blogs.wsj.com/indiarealtime/2011/09/20/agriculture-journal-china-versus-india-by-the-numbers/
70. Diane Toomey, 'How to Make Farm-to-Table a Truly Sustainable Movement', *Yale Environment 360,* 15 September 2014, http://e360.yale.edu/feature/interview_dan_barber_how_to_make_farm-to-table_a_truly_sustainable_movement/2803/#.VBr8wDHVmDY.twitter
71. Wendell Berry, *What Are People For?: Essays* (New York: North Point Press, 1990), 123.
72. Swaminathan, *From Green to Evergreen Revolution,* 23.
73. http://www.dnaindia.com/india/report-punjab-s-land-starved-farmers-head-for-australia-1141534
74. Vinod Jyani and family, interview with the author, 3–4 October 2012.
75. Subhash Palekar, *The Philosophy of Spiritual Farming* (Amravati, Maharashtra: Zero Budget Spiritual Farming, n.d.)
76. Ashish Ahuja, interview with the author, 1 October 2012.
77. Paul Christensen, 'ETC and the Story of the 'Cremate Monsanto' Operation in India: Part II', Seed in Context blog, 18 March 2013, http://www.intlcorn.com/seedsiteblog/?p=1241
78. Conway, *One Billion Hungry* ..., 250, citing Handa, B., 'Effect on fertilizer on groundwater quality in India', Groundwater in Resources Planning II: 1105–1109.
79. Nameet MVS, interview with the author, 26 November 2013.
80. Times News Network, 'Food Worth Rs 58k Crore Goes Waste in India Every Year', *Times of India,* 6 June 2013, http://

timesofindia. indiatimes.com/city/chennai/Food-worth-Rs-58k-crore-goes-waste-in-India-every-year/articleshow/20452251.cms

81. Catherine Greene, 'Growth Patterns in the US Organic Industry', US Department of Agriculture, 24 October 2103, http://www.ers.usda.gov/amber-waves/2013-october/growth-patterns-in-the-us-organic-industry.aspx#.VMZkLcbv8Z0
82. Swaminathan, *From Green to Evergreen Revolution*, 118.
83. Vinod K. Bhatt, interview with the author, 1 November 2013.
84. Rosegrant et al., 'Food Security in a World of Natural Resource Scarcity,' 9–10.
85. Swaminathan, *From Green to Evergreen Revolution*, 20.
86. C. Ford Runge, Benjamin Senauer, Philip G. Pardey and Mark W. Rosegrant, *Ending Hunger in our Lifetime: Food Security and Globalization* (Baltimore, MD and London, International Food Policy Research Institute and John Hopkins University Press, 8 July 2003), 15, citing Hopper, G.R., 'Changing Food Production and Quality of Diet in India: 1947–98', 1999, Population and Development Review (25)3: 443–77. Measham, A.R., and M. Chatterjee, 'Wasting Away: The Crisis in Malnutrition in India', Washington D.C.: World Bank.
87. Shawn Pogatchnik, '61.7 million Indian Children "Stunted": UN Report', *The New Indian Express,* 16 April 2013, http://www.newindianexpress.com/nation/61.7-million-Indian-children-stunted-UN-report/2013/04/16/article1547319.ece
88. Gardiner Harris, 'Study Says Pregnant Women in India Are Gravely Underweight', *New York Times,* 2 March 2015, http://www.nytimes.com/2015/03/03/world/asia/-pregnant-women-india-dangerously-underweight-study.html
89. Gardiner Harris, 'Poor Sanitation in India May Afflict Well-Fed Children with Malnutrition', *New York Times,* 13 July 2014, http://www.nytimes.com/2014/07/15/world/asia/poor-sanitation-in-india-may-afflict-well-fed-children-with-mal nutrition.html
90. Economic and Social Development Department, FAO, 'Introduction', in 'Reducing Poverty and Hunger: The Critical Role of Financing for Food, Agriculture and Rural Development', prepared for the International Conference on Financing for

Development, February 2002, http://www.fao.org/docrep/003/y6265e/y6265e03.htm

91. Rodney Fink, 'Corruption and the Agricultural Sector', Management Systems International, November 2002.
92. See The Long-Term Ecological Research Network, http://www.lternet.edu/
93. Charles, 'Fertilized World', 110.
94. Rosegrant et al., 'Food Security in a World of Natural Resource Scarcity', 14, citing de Ponti, T., B. Rijik and M.K. van Ittersum, 'The Crop Yield Gap between Organic and Conventional Agriculture', 2012, Agricultural Systems, 108: 1–9. Seufert. V.N., Ramakutty and J.A. Foley, 'Comparing the Yield of Organic and Conventional Agriculture, 2012, Nature, 485, 229–234.
95. Ann Arbor Science & Skeptics, 'Dr Catherine Badgley "The Sustainability and Efficacy of Organic Farming",' audio recording, Ann Arbor Science & Skeptics, 18 February 2102, http://annarborscienceskeptic.com/2012/podcast-2/download-now-dr-catherine-badgley-the-sustainability-efficacy-of-organic-farming/
96. Kari Hamershlag, 'Climate and Environmental Impacts', in *A Meat Eater's Guide to Climate Change and Health,* Environmental Working Group, July 2011, http://www.ewg.org/meateaters guide/a-meat-eaters-guide-to-climate-change-health-what-you-eat-matters/climate-and-environmental-impacts/, accessed on 6 March 2015.
97. Mark Bittman, 'How to Feed the World,' *New York Times,* 14 October 2013, http://www.nytimes.com/2013/10/15/opinion/how-to-feed-the-world.html?ref=international&utm_content=bufferd78ba&utm_source=buffer&utm_medium=twitter&utm_campaign=Buffer&_r=0
98. Tracie McMillan, 'The Future of Food: Field Studies', *OnEarth,* 7 July 2014, http://www.onearth.org/articles/2014/07/whats-the-state-of-the-american-farm-the-usda-tackles-issues-of-growing-concern
99. Bittman, 'How to Feed the World.'
100. Chetan Chauhan, 'India Wastes More Farm Food Than China: UN', *Hindustan Times,* 11 September 2013, http://www.

hindustantimes.com/newdelhi/india-wastes-more-farm-food-than-china-un/article1-1120755.aspx

101. 'Wasted: How America is Losing Up to 40 Per cent of Its Food from Farm to Fork to Landfill', Natural Resources Defense Council, 21 August 2012, http://www.nrdc.org/food/wasted-food.asp
102. Raphael Minder, 'Tempting Europe with Ugly Fruit', *New York Times,* 24 May 2014, http://www.nytimes.com/2014/05/25/world/europe/tempting-europe-with-ugly-fruit.html
103. FAO, 'Meeting the Food Security Challenge Through Organic Agriculture', FAONewsroom, 3 May 2007, http://www.fao.org/newsroom/en/news/2007/1000550/index.html
104. Ariel Bleicher, 'Farming by the Numbers', *IEEE Spectrum,* June 2013, 41–42.
105. Ibid., 44.
106. Lang, 'Seven-Year Glitch', and Donnelle Eller, '"Superweeds" Choke Farms', *Des Moines Register,* 23 June 2014, http://www.desmoinesregister.com/story/money/agriculture/2014/06/22/superweeds-choke-farms/11231231/
107. Rachel Carson, *Silent Spring* (New York: First Mariner Books, 2002), 245.
108. John Elliott, *Implosion: India's Tryst with Reality* (Noida, Uttar Pradesh: HarperCollins India, 2014), 64.
109. Agroecology in Action, 'What Is Agroecology', Agroecology in Action, http://agroeco.org/, accessed on 6 March 2015.
110. Sir Albert Howard, *An Agricultural Testament* (New York: Oxford University Press, 1943), 20.
111. 'Agriculture at a Crossroads: Global Report', in *International Assessment of Agricultural Knowledge, Science and Technology for Development,* (eds) Beverly D. McIntyre, Hans R. Herren, Judi Wakhungu, and Robert T. Watson, 2009, 23–34, http://www.unep.org/dewa/assessments/ecosystems/iaastd/tabid/105853/default.aspx
112. Judith D. Schwartz, 'Soil as Carbon Storehouse: New Weapon in Climate Fight?', *Yale Environment 360,* 4 March 2014, http://e360.yale.edu/feature/soil_as_carbon_storehouse_new_weapon_in_climate_fight/2744/

113. Ibid.
114. Todd A. Onti and Lisa A. Schulte, 'Soil Carbon Storage', *Nature Education Knowledge* 3, No. 10 (2012), 35, http://www.nature.com/scitable/knowledge/library/soil-carbon-storage-84223790
115. G. Philip Robertson, 'Soil Carbon,' The Long Term Ecological Research Network, accessed March 6, 2015, http://www.lternet.edu/node/83500.
116. National Science Foundation (NSF), 'Scientists Develop New Carbon Accounting Method to Reduce Farmers' Use of Nitrogen Fertilizer', NSF, 18 July 2012, http://www.nsf.gov/ news/news_summ.jsp?cntn_id=123848, accessed on 6 March 2015.
117. Paul Neate, 'Meta-Analysis of Crop Simulations Highlights Need for Adaptation Measures', Future Earth, 16 June 2014, http://www.futureearth.info/news/meta-analysis-crop-simulations-highlights-need-adaptation-measures, accessed on 6 March 2015; and A.J. Challinor, J. Watson, D.B. Lobell, S.M. Howden, D.R. Smith and N. Chhetri, 'A Meta-Analysis of Crop Yield Under Climate Change and Adaptation,' *Nature Climate Change* 4 (2014), 287–91.
118. Kay McDonald, 'An Interview with Cornell's Dr Erika Styger About the System of Crop Intensification (SRI-Rice)', Big Picture Agriculture, http://www.bigpictureagriculture.com/2013/07/interview-cornell-erika-styger-system-rice-intensification-sri-408.html, accessed on 6 March 2015.
119. SRI-Rice, 'Madagascar', SRI International Network and Resources Center (SRI-Rice), Cornell University, College of Agriculture and Life Sciences, http://sri.ciifad.cornell.edu/countries/madagascar/, accessed on 6 March 2015.
120. SRI-Rice, 'SCI-The System of Crop Intensification: Agroecological Innovations for Improving Agricultural Production, Food Security, and Resilience to Climate Change', SRI International Network and Resources Center (SRI-Rice), Cornell University and the Technical Centre for Agricultural and Rural Cooperation (CTA), 2014.
121. John Vidal, 'Miracle Grow: Indian Rice Farmer Uses Controversial Method for Record Crop', *The Guardian,* 12 May 2014, http://www.theguardian.com/global-development/2014/

may/13/miracle-grow-indian-rice-farmer-sri-system-rice-intensification-record-crop

122. Raksha Kumar, 'For Farmers in Bihar, a Simple Solution for More Crops', *New York Times,* 9 April 2013, http://india.blogs.nytimes.com/2013/04/09/for-farmers-in-bihar-a-simple-solution-for-more-crops/
123. Olivier De Schutter, 'Agroecology and the Right to Food', report presented at the 16th Session of the United Nations Human Rights Council [A/HRC/16/49], 20 December 2010, 8, http://www.srfood.org/en/official-reports
124. Devinder Sharma, 'Not Crop Diversification, Punjab Needs to Diversify from Existing Intensive Farming System', Ground Reality blog, http://devinder-sharma.blogspot.com/2013/01/not-crop-diversification-punjab-needs.html, accessed on 6 March 2015.
125. Ibid.
126. Kumar, 'For Farmers in Bihar.'
127. Vidal, 'Miracle Grow.'
128. 'Rice-to-Riches Story for Paddy Cultivators in Dangs', *DNA*, 24 May 2014, http://www.dnaindia.com/ahmedabad/report-rice-to-riches-story-for-paddy-cultivators-in-dangs-1991034, accessed on 6 March 2015.
129. Vidal, 'Miracle Grow.'
130. Chander Suta Dogra, 'Punjab's New Agro Policy Will Be a Drain on Hope', *The Hindu,* 28 July 2013, http://www.thehindu.com/news/national/other-states/punjabs-new-agro-policy-will-be-a-drain-on-hope/article4960714.ece
131. Moudgil, 'Punjab Wades in Shallow Waters.'
132. http://www.thehindu.com/data/does-it-pay-to-be-a-farmer-in-india/article6713980.ece
133. Rosegrant et al., 'Food Security in a World of Natural Resource Scarcity', 15, citing Willer 2011.
134. Laura Reynolds and Catherine Ward, 'Certified Organic Farmland Still Lagging Worldwide', Worldwatch Institute, 15 January 2013, http://www.worldwatch.org/certified-organic-farmland-still-lagging-worldwide, accessed on 6 March 2015.

2. AP—WATER

1. Fakir Mohan Senapati, 'Asura Pond', in Chandrahas Choudhury, (ed.), *India: A Traveller's Literary Companion* (New Delhi: HarperCollins, 2010), 106.
2. W.H. Auden, *Collected Poems* (New York: Modern Library, 2007), 582.
3. Water for Asian Cities, 'How Delhi Makes the Sprightly Yamuna a "Dead River",' Water for Asian Cities, http://www.unwac.org/research_articles.php, accessed on 5 March 2015.
4. IBN7, *IBNLive*, 7 June 2013, http://ibnlive.in.com/news/rajasthan-many-cities-get-water-every-4th-day-due-to-shortage/396982–3-239.html
5. Deepak Malik, 'Without Rain, A Bleak Outlook,' India Together, October 2002, http://indiatogether.org/environment/articles/droughtrn.htm, accessed on 5 Marc 2015.
6. Kanhaiya Lal Gurjar, interview with author, 20 November 2012.
7. A. Vaidyanathan, *Water Resources of India* (New Delhi: Oxford University Press, 2013), xxxvi–xxxvii.
8. United Nations, 'Water Scarcity', International Decade for Action 'Water for Life' 2005–2015, http://www.un.org/waterforlifedecade/scarcity.shtml
9. Ibid.
10. Mark Fischetti, 'How Much Water Do Nations Consume?' *Scientific American,* 21 May 2012, http://www.scientificamerican.com/article/graphic-science-how-much-water-nations-consume/
11. Vaidyanathan, *Water Resources of India,* 8.
12. Ibid.
13. Lester R. Brown, 'Stabilizing Water Tables: Falling Water Tables', Chapter 6, in *Outgrowing the Earth: The Food Security Challenge in an Age of Falling Water Tables and Rising Temperatures* (Washington DC: Earth Policy Books, 2004), http://www.earth-policy.org/books/out/ote6_2
14. Vaidyanathan, *Water Resources of India,* xxxiii.
15. 'Managing Water Under Uncertainty and Risk', *The United Nations World Water Development Report 4,* Volume 1 (Paris: United Nations Educational, Scientific and Cultural Organization, 2012), 5, http://www.unesco.org/new/en/natural-sciences/environment/water/wwap/wwdr/wwdr4-2012/

16. Vaidyanathan, *Water Resources of India,* xxxii, xxxvii.
17. Brian Murton, 'Famine', in *The Cambridge World History of Food,* Vol. 1 (New York: Cambridge University Press, 2000), 1412; and Amartya Sen, *Poverty and Famines: An Essay on Entitlement and Deprivation* (London: Oxford University Press, 1981), 203.
18. Vaidyanathan, *Water Resources of India,* 38–40.
19. Ibid., 52.
20. Ibid., 79–80.
21. Diana L. Eck, *Banaras: City of Light* (Princeton, NJ: Princeton University Press, 1982), 29.
22. 'Hirakud Dam', *Wikipedia,* http://en.wikipedia.org/wiki/Hirakud_Dam, accessed on 5 March 2015.
23. Vaidyanathan, *Water Resources of India,* xxiv.
24. Cheryl Colopy, *Dirty, Sacred Rivers: Confronting South Asia's Water Crisis* (New York: Oxford University Press), 141–46.
25. Vaidyanathan, *Water Resources of India,* xxv; and 'World's Oldest Dam—Grand Anicut Dam', *Sanskriti,* 8 January 2014, http://www.sanskritimagazine.com/history/worlds-oldest-dam-grand-anicut-dam/#, accessed on 4 March 2015.
26. Vaidyanathan, *Water Resources of India,* 3.
27. Colopy, *Dirty Sacred Rivers,* 142.
28. United Nations Children's Fund (UNICEF), 'Take Poo to the Loo', UNICEF, http://www.poo2loo.com/fact-and-stats.php, accessed on 5 March 2015.
29. Colopy, *Dirty Sacred Rivers,* 61.
30. Ibid., 63
31. Vaidyanathan, *Water Resources of India,* 19.
32. Ibid., xxv.
33. Jacques Leslie, *Deep Water: The Epic Struggle over Dams, Displaced People, and the Environment* (New York: Picador, 2005), 44.
34. Vaidyanathan, *Water Resources of India,* 28.
35. Leslie, *Deep Water,* 43.
36. Vaidyanathan, *Water Resources of India,* 83.
37. Ibid., 84.
38. Ibid., 85.
39. Atif Ansar et al., 'Should We Build More Large Dams? The Actual Costs of Hydropower Megaproject Development', *Energy Policy,* March, 2014.

40. Amy Kober, 'Elwha River's Health Rebounding as Biggest Dam Removal in History Nears Completion', American Rivers, 25 August 2014, http://www.americanrivers.org/newsroom/press-releases/elwha-rivers-health-rebounding/, accessed on 5 March 2015.
41. Leslie, *Deep Water,* 5.
42. Ibid., 24.
43. Madhav Gadgil and Ramachandra Guha, *The Use and Abuse of Nature: Incorporating This Fissured Land: An Ecological History of India and Ecology and Equity* (New Delhi: Oxford University Press, 2000), 4; and Leslie, *Deep Water,* 60.
44. World Bank, *Environmental Assessment Sourcebook: Guidelines for Environmental Assessment of Energy and Industry Project,* World Bank Publications, 1992, 86.
45. Muhammad Akbar Notezai, 'Interview: The India–Pakistan Water Dispute', The Diplomat, 21 November 2014, http://thediplomat.com/2014/11/interview-the-india-pakistan-water-dispute/, accessed on 5 March 2015.
46. Hartosh Singh Bal, *Waters Close Over Us: A Journey Along the Narmada* (Noida: Fourth Estate, 2013), 143–44.
47. Vaidyanathan, *Water Resources of India,* 9.
48. Colopy, *Dirty Sacred Rivers,* 266.
49. Navin Singh Khadka, 'Concerns over India Rivers Order', BBC News, 30 March 2012, http://www.bbc.com/news/science-environment-17555918, accessed on 5 March 2015.
50. 'Fast Track Irrigation Projects, Identify Rivers for Linking: PM Modi', *ZeeNews,* 30 December 2014, http://zeenews.india.com/news/india/fast-track-irrigation-projects-identify-rivers-for-linking-pm-modi_1522525.html, accessed on 5 March 2015.
51. Ellen Barry and Neha Thirani Bagri, 'Narendra Modi, Favoring Growth in India, Pares Back Environmental Rules', *New York Times,* 4 December 2014, http://www.nytimes.com/2014/12/05/world/indian-leader-favoring-growth-sweeps-away-environmental-rules.html?_r=0, accessed on 5 March 2015.
52. Vaidyanathan, *Water Resources of India,* 36.
53. Ibid., 26.
54. Rajendra Singh, interviews with the author, 16–24 November 2012.

55. Dhanna Gurjar, interview with the author, 24 November 2012.
56. Tarun Bharat Sangh, 'River Arvari Parliament', Tarun Bharat Sangh, http://tarunbharatsangh.in/river-arvari-parliament/, accessed on 5 March 2015.
57. Ibid.
58. Vaidyanathan, *Water Resources of India,* 96.
59. Nitya Jacob, *Jalyatra: Exploring India's Traditional Water Management Systems* (New Delhi: Penguin Books, 2008), 257.
60. P. Sakthivel, S. Amirthalingam and M. Starkl, 'A Study on Law Relating to Groundwater Recharge in India', *Rostrum's Law Review,* Volume 1, Issue IV, 30 July 2014, http://rostrumlegal.com/blog/a-study-on-law-relating-to-groundwater-recharge-in-india-by-p-sakthivel-s-amirthalingam-m-starkl/, accessed on 5 March 2015.
61. Ramesh Menon, 'One Village. 60 Millionaires. The Miracle of Hiware Bazar', *Tehelka* 42, No. 9 (20 October 2012), http://www.tehelka.com/one-village-60-millionaires-the-miracle-of-hiware-bazar/
62. Ibid.
63. Civil Society Online, 'Rocky Perla Grows a Forest', Civil Society Online, December 2012, http://www.civilsocietyonline.com/pages/Details.aspx?212, accessed on 5 March 2015.
64. http://www.hindustantimes.com/india-news/world environment/green-heroes-odisha-s-conservation-master/article1–1070897.aspx.
65. Jacob, *Jalyatra,* 41–69.
66. 'Mega Dream: Rs 22,000 Cr to Restore 45,300 Minor Irrigation Tanks', *The New Indian Express,* 26 September 2014, http://www.newindianexpress.com/states/telangana/Mega-Dream-Rs-22000-Cr-to-Restore-45300-Minor-Irrigation-Tanks/2014/09/26/article2449847.ece, accessed on 5 March 2015.
67. Ker Than, 'Artificial Glaciers Water Crops in Indian Highlands', *National Geographic,* 14 February 2012, http://news.nationalgeographic.com/news/2012/02/120214-artificial-glaciers-water-crops-in-indian-highlands/, accessed on 5 March 2015.
68. 'A Himalayan Village Builds Artificial Glaciers to Survive Global

Warming [Slide Show]', *Scientific American*, http://www.scientificamerican.com/slideshow/artificial-glaciers-to-survive-global-warming/, accessed on 5 March 2015.

69. Preeti Mohan, 'I Started to Do RWH in Chennai for Very Selfish Reasons'—Sekhar Raghavan, Rain Centre', The Alternative, http://www.thealternative.in/society/i-started-to-do-rwh-in-chennai-for-selfish-reasons-sekhar-raghavan/, accessed on 11 March 2013.
70. 'Toxins, Bacteria in Groundwater', *The Times of India*, 22 August 2103, http://timesofindia. indiatimes.com/city/kolkata/Toxins-bacteria-in-groundwater/articleshow/21968444.cms, accessed on 5 March 2015.
71. Claire Jean Glendenning, interview with the author, 4 December 2012.
72. Claire Jean Glendenning, 'Evaluating the Impacts of Rainwater Harvesting in a Case Study Catchment: The Arvari River, Rajasthan, India', University of Sydney doctoral dissertation, 29 August 2009.
73. Notably: M. Dinesh Kumar, executive director of Institute for Resource Analysis and Policy in Hyderabad and co-editor of *The Water, Energy and Food Security Nexus: Lessons from India for Development.*
74. M. Dinesh Kumar, Nitin Bassi, A. Narayanamoorthy and M.V.K. Sivamohan (eds), *The Water, Energy and Food Security Nexus: Lessons from India for Development* (New York: Routledge, 2014), 8.
75. Leora Broydo Vestel, 'The Legalities of Rainwater Harvesting', *New York Times*, 29 June 2009, http://green.blogs.nytimes.com/2009/06/29/the-legalities-of-rainwater-harvesting/?_r=0, accessed on 5 March 2015.
76. Glendenning was able to assess that it was the same flow by testing the electrical conductivity.
77. Glendenning, ' Evaluating the Impacts of Rainwater Harvesting', 166.
78. David L. Haberman, *River of Love in an Age of Pollution: The Yamuna River of Northern India* (Berkeley: University of California Press, 2006), 102.

79. Gabriel Mante and Mattias Kolstrup, 'Environmental Refugees—How Climate Change Affects People's Lives', RESET, August 2014, http://en.reset.org/knowledge/environmental-refugees-%E2%80%93-how-climate-change-affects-peoples-lives, accessed on 5 March 2015.
80. Rachel Nuwer, 'An Ancient Civilization, Upended by Climate Change', *New York Times,* 29 May 2012, http://green.blogs.nytimes.com/2012/05/29/an-ancient-civilization-upended-by-climate-change/?ref=energy-environment, accessed on 5 March 2015.
81. Mayank Vahia, 'Did the River Saraswati, Mentioned in the Vedas, Really Exist?', *DNA,* 3 October 2014, http://www.dnaindia.com/analysis/standpoint-did-the-river-saraswati-mentioned-in-the-vedas-really-exist-2023285.
82. Kanhaiya Lal Gurjar, Rama Devi and Lada, interview with the author, 24 November 2012.
83. http://www.theroyalforums.com/forums/f166/prince-charles-and-camilla-parker-bowles-2-july-nov-2003-a-1150–2.html.
84. International River Foundation, 'Thiess International Riverprize,' International RiverFoundation website, http://www.riverfoundation.org.au/riverprize_international.php, accessed on 5 March 2015.
85. Vaidyanathan, *Water Resources of India,* 1.

3. AGNI—FIRE

1. Lao-tzu, *Tao Te Ching,* translated by Stephen Mitchell (New York: HarperCollins, 2009), verse 79, http://acc6.its.brooklyn.cuny.edu/~phalsall/texts/taote-v3.html
2. Seema Dattabay Kolekar, interview with the author, 25 December 2013.
3. Dr C. Chandramouli, 'Housing, Household Amenities and Assets—Key Results from Census 2011', Census of India, Government of India, 2012, slide 57 and 58, http://www.censusindia.gov.in/2011census/hlo/Data_sheet/India/00_2011_Housing_India.ppt. 160 million figure from: Gautam N. Yadama, *Fires, Fuel & the Fate of 3 Billion: The State of the Energy Impoverished* (New York: Oxford University Press, 2013), 40,

Venkataraman, C. et al., 'The Indian National Initiative for Advanced Biomass Cookstoves: The Benefits of Clean Combustion', Energy for Sustainable Development, 2010. Kim, Stephen S., et al., 'A Comparative Risk Assessment of Burden of Disease and Injury Attributable to 67 Risk Factors and Risk Factor clusters in 21 Regions: 1990–2010: A Systematic Analysis for the Global Burden of Disease Study 2010', Lancet, 2012. Vos, T. et al., 'Years lived with a Disability (YLDs) for 1160 Sequelae of 289 Diseases and Injuries: 1990–2010: Systematic Analysis for the Global Burden of Disease Study 2010', Lancet, 2012. Wang, H. et al., 'Age-specific and Sex-specific Mortality in 187 Countries: 1970–2010: A Systematic Analysis for the Global Burden of Disease Study, Lancet, 2012.

4. Eighty-seven per cent of homes are dependent on biomass. Office of the Registrar General and Census Commissioner India, 'Houselisting and Housing Census Data Highlights—2011', Census of India, Government of India, 2012, http://www.censusindia.gov.in/2011census/hlo/hlo_highlights.html; and Dr C. Chandramouli, 'Housing, Household Amenities and Assets—Key Results from Census 2011', Census of India, Government of India, 2012, slide 57 and 58, http://www.censusindia.gov.in/2011census/hlo/Data_sheet/India/00_2011_Housing_India.ppt.
5. Kirk R. Smith, interview with the author, 4 October 2012, and subsequent email and telephone correspondence.
6. K.R. Smith, N. Bruce, K. Balakrishnan, H. Adair-Rohani, J. Balmes, Z. Chafe, M. Dherani, H.D. Hosgood, S. Mehta, D. Pope and E. Rehfuess, 'Millions Dead: How Do We Know and What Does It Mean? Methods Used in the Comparative Risk Assessment of Household Air Pollution', *Annual Review of Public Health* 35 (2014), 185–206.
7. World Health Organization (WHO), 'Household Air Pollution and Health', WHO Fact Sheet No. 292, March 2014, http://www.who.int/mediacentre/factsheets/fs292/en/; and Smith et al., 'Millions Dead'.
8. Kalpana Balakrishnan, Aaron Cohen and Kirk R. Smith, 'Addressing the Burden of Disease Attributable to Air Pollution

in India: The Need to Integrate Across Household and Ambient Air Pollution Exposures', *Environmental Health Perspective* 122, No. 1 (2014), doi:10.1289/ehp.1307822

9. David Biello, 'Brown Haze from Cooking Fires Cooking Earth, Too', *Scientific American,* 1 August 2007, http://www.scientificamerican.com/article.cfm?id=brown-haze-from-cooking-fires-cooking-earth
10. Global Alliance for Clean Cookstoves, '2012 Results Report: Sharing Progress on the Path to Adoption of Clean Cooking Solutions', iii, 25 September 2013, http://cleancookstoves.org/resources/221.html, accessed on 18 March 2015.
11. Global Alliance for Clean Cookstoves, 'Carbon Credit Prices for Improved Cookstove Projects', Global Alliance for Clean Cookstoves, http://carbonfinancefor cookstoves.org/carbon-finance/prices-for-improved-cookstove-projects/, accessed on 6 March 2015.
12. Vanessa Able, 'Tata Nano, the Car that Was Just Too Cheap', *The Guardian,* 3 February 2014, http://www.theguardian.com/commentisfree/2014/feb/03/tata-nano-car-cheap-poor-safety-rating
13. Kalpana Balakrishnan, interview with the author, 10 March 2014.
14. Yadama, *Fires, Fuel & the Fate of 3 Billion,* 75.
15. Vanita Kolekar, interview with the author, 23 December 2013.
16. Radha Muthiah, '4 Million Reasons for Clean Cookstoves', Global Connections, UN Foundation Blog, 20 October 2014, http://unfoundationblog.org/4-million-reasons-for-clean-cookstoves/#sthash.SgNSlFWT.dpuf, accessed on 6 March 2015.
17. US Department of State, 'Global Alliance for Clean Cookstoves', US Department of State, http://www.state.gov/s/partnerships/cleancookstoves/, accessed on 6 March 2015.
18. Kirk R. Smith, e-mail correspondence with the author, 13 November 2013.
19. Malaria = 627,000 deaths in 2012: WHO, 'Malaria', WHO Fact Sheet no. 94, December 2014, http://www.who.int/mediacentre/factsheets/fs094/en/. TB = one million deaths in 2011: WHO, 'The Top 10 Causes of Death,' WHO Fact Sheet

No. 310, accessed March 6, 2015, http://www.who.int/mediacentre/factsheets/fs310/en/index2.html. HIV/AIDS = 1.7 million in 2012: WHO, 'Number of Deaths Due to HIV/AIDS Estimates by WHO Region,' WHO, accessed March 6, 2015, http://apps.who.int/gho/data/view.main.22600WHO?lang=en, accessed on 6 March 2015.

20. Fen Montaigne, 'Record 400 ppm CO_2 Milestone "Feels Like We're Moving into Another Era",' *The Guardian,* 14 May 2013, http://www.theguardian.com/environment/2013/may/14/record-400ppm-co2-carbon-emissions
21. M.B. Epstein, M.N. Bates, N.K. Arora, K. Balakrishnan, D.W. Jack and K.R. Smith, 'Household Fuels, Low Birth Weight, and Neonatal Death in India: The Separate Impacts of Biomass, Kerosene, and Coal', *International Journal of Hygiene and Environmental Health* 216 (2012), 523–32.
22. Agency for Toxic Substances & Disease Registry, 'ToxFAQs for Formaldehyde', Agency for Toxic Substances & Disease Registry, September 2008, http://www.atsdr.cdc.gov/toxfaqs/tf.asp?id=219&tid=39, accessed on 6 March 2015.
23. Epstein et al., 'Household Fuels, Low Birth Weight, and Neonatal Death in India.'
24. Burkhard Bilger, 'Hearth Surgery', *The New Yorker,* 21 December 2009, 84. http://archives.newyorker.com/?i=2009-12-21#folio=084; and Yadama, *Fires, Fuel & the Fate of 3 Billion,* 46.
25. Agency for Toxic Substances & Disease Registry, 'ToxFAQs for Chloromethane', Agency for Toxic Substances & Disease Registry, June 1999, http://www.atsdr.cdc.gov/toxfaqs/tf.asp?id=586&tid=109, accessed on 6 March 2015.
26. Kalpana Balakrishnan, Jyoti Parikh, Sambandam Sanka, Ramaswamy Padmavathi, Kailasam Srividya, Vidya Venugopal, Swarna Prasad and Vijay Laxmi Pandey, 'Daily Average Exposures to Respirable Particulate Matter from Combustion of Biomass Fuels in Rural Households of Southern India', *Environmental Health Perspectives,* 110, No. 11, November 2002, 1069.
27. Epstein et al., 'Household Fuels, Low Birth Weight, and Neonatal Death in India.'

28. The skin, liver, and immune system: Agency for Toxic Substances & Disease Registry, 'Polycyclic Aromatic Hydrocarbons (PAHs)', Agency for Toxic Substances & Disease Registry, last updated 3 March 2011, http://www.atsdr.cdc.gov/substances/toxsubstance.asp?toxid=25, accessed on 6 March 2015.
29. Balakrishnan et al., 'Daily Average Exposures to Respirable Particulate Matter'.
30. Agency for Toxic Substances & Disease Registry, 'ToxFAQs for Nitrogen Oxides,' Agency for Toxic Substances & Disease Registry, April 2002, http://www.atsdr.cdc.gov/toxfaqs/tf.asp?id=396&tid=69, accessed on 6 March 2015.
31. Epstein et al., 'Household Fuels, Low Birth Weight, and Neonatal Death in India.'
32. Union of Concerned Scientists, 'Does Air Pollution—Specifically Particulate Matter (Aerosols)—Affect Global Warming?', Union of Concerned Scientists, http://www.ucsusa.org/global_warming/science_and_impacts/science/aerosols-and-global-warming-faq.html, accessed on 6 March 2015.
33. WHO, 'Ambient (Outdoor) Air Quality and Health', WHO Fact Sheet N0. 313, March 2014, http://www.who.int/mediacentre/factsheets/fs313/en/, accessed on 6 March 2015.
34. Elisabeth Rosenthal, 'Third-World Stove Soot Is Target in Climate Fight', *New York Times,* 15 April 2009, http://www.nytimes.com/2009/04/16/science/earth/16degrees.html?pagewanted=all&_r=0
35. Kathleen H. Harriman and Lisa M. Brosseau, 'Controversy: Respiratory Protection for Healthcare Workers', *Medscape*, 28 April 2011.
36. 'Ambient (and Outdoor) Air Quality and Health', Fact sheet No. 313, World Health Organization, March 2014, http://www.who.int/mediacentre/factsheets/fs313/en/, accessed on 19 March 2015.
37. S.S. Lim et al., 'A Comparative Risk Assessment of Burden of Disease and Injury'.
38. WHO, 'Household Air Pollution and Health', WHO Fact Sheet No. 292, March 2014, http://www.who.int/mediacentre/factsheets/fs292/en/, accessed on 6 March 2015.

39. *India Clean Cookstove Forum 2013* (New Delhi: Deutsche Gesellschaft für Internationale Zusammenarbeit GmbH or GIZ, 2013), 1, https://energypedia.info/wiki/File:Report_on_Outcomes_India_Clean_Cookstove_Forum_2013.pdf.
40. 10 μg/m^3 compared to 337 μg/m^3, from Smith et al., 'Millions Dead', 191; Smith phone interview with author, 13 May 2014; and WHO, 'Household Air Pollution and Health', WHO Fact Sheet No. 292, March 2014, http://www.who.int/mediacentre/factsheets/fs292/en/, accessed on 6 March 2015.
41. Rema Hanna, Esther Duflo and Michael Greenstone, 'Up in Smoke: The Influence of Household Behavior on the Long-Run Impact of Improved Cooking Stoves', NBER Working Paper No. 18033, May 2012, http://www.nber.org/papers/w18033
42. Narendra Zende, interview with the author, 23 December 2013.
43. Appropriate Rural Technology Institute, 'Briquetted Charcoal from Sugarcane Trash', Appropriate Rural Technology Institute, http://www.arti-india.org/index.php?option=com_content&view=article&id=42:briquetted-charcoal-from-sugarcane-trash&catid=15:rural-energy-technologies&Itemid=52, accessed on 6 March 2015.
44. Lata Kisan Khare, interview with the author, 23 December 2013.
45. US Department of State, 'Global Alliance for Clean Cookstoves',
46. Gardiner Harris, 'Beijing's Bad Air Would Be Step Up for Smoggy Delhi', *New York Times,* 24 January 2014, http://www.nytimes.com/2014/01/26/world/asia/beijings-air-would-be-step-up-for-smoggy-delhi.html; and Environmental Performance Index, 'China-India Smog Rivalry a Sign of Global Menace', *The Metric,* 27 March 2014, http://epi.yale.edu/the-metric/china-india-smog-rivalry-sign-global-menace
47. Smith et al., 'Millions Dead'. 192.
48. Kalpana Balakrishnan, interview with the author, 10 March 2014.
49. Biello, 'Brown Haze from Cooking Fires.'
50. Regina Nuzzo, 'Biography of Veerabhadran Ramanathan', *Proceedings of the National Academy of Sciences of the United States,* 102, No. 15, 12 April 2005, 5323–25.

51. Ibid.
52. Azeen Ghorayshi, 'India Air Pollution Cutting Crop Yields by Almost Half', *The Guardian,* 3 November 2014, http://www.theguardian.com/environment/2014/nov/03/india-air-pollution-cutting-crop-yields-by-almost-half
53. 'Atmospheric Brown Clouds: Regional Assessment Report with Focus on Asia', United Nations Environment Programme, (2008), 5, http://www.unep.org/documents.multilingual/default.asp?DocumentID=550&ArticleID=5978&l=en
54. Lester R. Brown, *Melting Mountain Glaciers Will Shrink Grain Harvests in China and India,* Earth Policy Institute, 2008, http://www.earth=policy.org/plan_b_updates/2008/update71, accessed on 22 August 2015.
55. Pam Pearson, phone interview with the author, 20 March 2014.
56. 'On Thin Ice: How Cutting Pollution Can Slow Warming and Save Lives' (Washington DC: World Bank and International Cryosphere Climate Initiative, 2013), p. 55.
57. T.C. Bond et al., 'Bounding the Role of Black Carbon in the Climate System: A Scientific Assessment', *Journal of Geophysical Research-Atmospheres,* 2013, doi:10.1002/jgrd.50171
58. Ibid., p. 5381–85.
59. Ibid., p. 5385.
60. Ibid., p. 5523.
61. Jessica Seddon Wallack and Veerabhadran Ramanathan, 'The Other Climate Changers: Why Black Carbon and Ozone Also Matter', *Foreign Affairs,* October 2009, http://www.foreignaffairs.com/articles/65238/jessica-seddon-wallack-and-veerabhadran-ramanathan/the-other-climate-changers
62. Carbon Brief, 'How Long Do Greenhouse Gases Stay in the Air?', *The Guardian*'s Ultimate Climate Change FAQ, http://www.theguardian.com/environment/2012/jan/16/greenhouse-gases-remain-air, accessed on 6 March 2015.
63. Union of Concerned Scientists, 'Does Air Pollution Affect Global Warming?'
64. Global Alliance for Clean Cookstoves, 'Partner Directory', Global Alliance for Clean Cookstoves, http://cleancookstoves.org/partners/directory.html

65. Global Alliance for Clean Cookstoves, '2012–2013 Developing Markets Worldwide', Global Alliance for Clean Cookstoves, 12.
66. Harish Anchan, interview with the author, 21 December 2014.
67. Ravi Kumar, 'Jivan Jyoti', YouTube video posted 18 July 2010, https://www.youtube.com/watch?v=eWi26rwYsd8
68. Mouhsine Serrar, interviews with the author, 25 and 29 January 2014.
69. Ted Nordhaus and Michael Shellenberger, 'Second Life', *The New Republic,* 24 September 2007, http://www.newrepublic.com/article/environment-energy/76102/second-life
70. Gautam Yadama, phone interview with the author, 21 March 2014.
71. Envirofit, 'Impact: Fighting Household Air Pollution', Envirofit, http://www.envirofit.org/india1/?sub=impact, accessed on 6 March 2015.
72. Fiona Lambe and Aaron Atteridge, 'Putting the Cook Before the Stove: A User-Centered Approach to Understanding Household Energy Decision-Making—A Case Study of Haryana State, Northern India,' Stockholm Environment Institute, 2012,7; and B. Sinha, 'The Indian Stove Programme: An Insider's View—the Role of Society, Politics, Economics, and Education,' *Boiling Point* 48 (2002).
73. Kirk R. Smith, A.L. Aggarwal and R.M. Dave, 'Air Pollution and Rural Biomass Fuels in Developing Countries: A Pilot Village Study in India and Implications for Research and Policy', *Atmospheric Environment* 17, No. 11 (1983), 2343–62.
74. K.R. Smith, Skype interview with the author, 13 May 2014.
75. Hanna et al, 'Up in Smoke: The Influence of Household Behavior on the Long-Run Impact of Improved Cooking Stoves'.
76. Rema Hanna, phone interview with the author, 4 September 2013.
77. Priyadarshini Karve, interview with the author, 26 December 2013.
78. Karabi Dutta, interview with the author, 1 December 2013.
79. Gireesh Shrimali, Xander Slaski, Mark C. Thurber and Hisham Zerriffi, 'Improved Stove in India: A Study of Sustainable Business Models', *Energy Policy,* 2011; and Douglas R. Barnes, Keith

Openshaw, Kirk R. Smith and Robert van der Plas, 'The Design and Diffusion of Improved Cooking Stoves', *The World Bank Research Observer*, 1993, http://www.jstor.org/discover/10.2307/3986529?uid=3738256&uid=2129&uid=2&uid=70&uid=4&sid=21103396727743

80. Atul Gawande, 'Slow Ideas: Some Innovations Spread Fast. How Do You Speed the Ones that Don't?' *The New Yorker*, 29 July 2013, http://www.newyorker.com/reporting/2013/07/29/130729fa_fact_gawande
81. http://www.dnaindia.com/money/report-tamil-nadu-number-one-state-on-development-parametersassocham-2068959
82. http://www.hardnewsmedia.com/2013/12/6156?page=2
83. Kirk R. Smith, 'National Burden of Disease in India from Indoor Air Pollution', *Proceedings of the National Academy of Sciences of the United States of America* 97, No. 24 (2000), 13286–93.
84. Bjorn Lomborg, 'The Poor Need Cheap Fossil Fuels', *New York Times*, 3 December 2013, http://www.nytimes.com/2013/12/04/opinion/the-poor-need-cheap-fossil-fuels.html?_r=0
85. Kirk R. Smith, 'In Praise of Petroleum?', *Science*, 6 December 2002, http://www.sciencemag.org/content/298/5600/1847.citation
86. 'Summary for Policymakers', Working Group III (WG III) of the Intergovernmental Panel on Climate Change (IPCC), 3rd volume of the IPCC's Fifth Assessment Report (AR5—*Climate Change 2013*), 32, http://www.citepa.org/en/news/1535-13-april-2014-climate-the-ipcc-releases-the-summary-for-policymakers-of-the-3rd-volume-of-its-5th-assessment-report
87. 'LPG Imports Cut by $1 Billion Backed by Government Measures', *Business Standard*, 9 October 2013, http://www.business-standard.com/article/economy-policy/lpg-imports-cut-by-1-billion-backed-by-government-measures-113100901172_1.html; and US Energy Information Administration (EIA), 'India—Analysis', US EIA, last updated 26 June 2014, http://www.eia.gov/countries/cab.cfm?fips=in, accessed on 6 March 2015.
88. Sumi Mehta, phone interview with the author, 27 March 2014.
89. International Energy Agency (IEA), 'World Energy Outlook,

Energy Subsidies', IEA, http://www.worldenergyoutlook.org/resources/energysubsidies/, accessed on 6 March 2015.

90. Brad Plumer, 'These 5 Charts Show Why the World Is Still Failing on Climate Change', *Vox,* 19 June 2014, http://www.vox.com/2014/6/19/5821250/these-5-charts-show-why-the-world-is-still-failing-on-climate-change
91. Jeff Tollefson, 'Energy: Islands of Light', *Nature,* 11 March 2014, http://www.nature.com/news/energy-islands-of-light-1.14860
92. Amy Yee, 'India Increases Effort to Harness Biomass Energy', *New York Times,* 7 October 2013, http://www.nytimes.com/2013/10/09/business/energy-environment/india-increases-effort-to-harness-biomass-energy.html?pagewanted=all&_r=0
93. Shruti Ravindran, 'India's Push for Renewable Energy: Is It Enough?', *National Geographic,* 19 September 2014, http://news.nationalgeographic.com/news/energy/2014/09/140919-india-modi-renewable-energy-science-world-wind-solar/
94. IEA, 'World Energy Outlook, Energy Subsidies'.

4. VAYU—AIR

1. K.N. Dave, *Birds in Sanskrit Literature: With 107 Bird Illustrations* (New Delhi: Motilal Banarsidass, 2005), 197.
2. Asad Rahmani, 'Race to Save Vultures', *Hornbill,* Oct–Dec 2008.
3. Vulture Conservation Foundation (VCF), 'Vultures', Vulture Conservation Foundation, http://www.4vultures.org/vultures/, accessed on 6 March 2015.
4. S.M. Satheesan, 'The More Serious Vulture Hits to Military Aircraft in India Between 1980 and 1994', Bird Strike Committee Europe, BSCE22/WP23, Vienna, 29 August–2 September 1994, http://www.academia.edu/5284775/BIRD_STRIKE_COMMITTEE_EUROPE_THE_MORE_SERIOUS_VULTURE_HITS_TO_MILITARY_AIRCRAFT_IN_INDIA_BETWEEN_1980_AND_1994
5. Amelia Gentleman, 'India's Vultures Fall Prey to a Drug in the Cattle They Feed On', *New York Times,* 28 March 2006, http://www.nytimes.com/2006/03/28/international/asia/28vultures.html?_r=0
6. D. Houston, 'Indian White-Backed Vulture *Gyps bengalensis*', in

I. Newton, and R.D. Chancellor (eds), *Conservation Studies on Raptors* (Cambridge, UK: International Council for Bird Preservation), Technical Publication No. 5 ICBP, 1985, 465–66.

7. Vibhu Prakash, phone interview with the author, 8 March 2008.
8. Lindsay Oaks, phone interview with the author, 7 March 2008.
9. Lindsay Oaks, 'Asian Vulture Crisis Notes from the Field: Veterinary Work on White-Backed Vultures in Pakistan', The Peregrine Fund website, 17 April 2001, http://blogs.peregrinefund.org/article/393, accessed on 6 March 2015.
10. Oaks, interview, 2008.
11. Munir Virani, interview with the author, 21 August 2006.
12. Rhys Green, I. Newton, S. Shultz, A. Cunningham, M. Gilbert, D.J. Pain and V. Prakash, 'Diclofenac Poisoning as a Cause of Vulture Population Declines Across the Indian Subcontinent', *Journal of Applied Ecology*, 41 (2004): 793–800, http://www.save-vultures.org/save_resources_scientificpapers.html
13. Lindsay Oaks, phone interview with the author, 21 April 2009.
14. Fifty-seven thousand of all *Gyps* figure from V. Prakash, R.E. Green, D.J. Pain, S.P. Ranade, S. Saravanan, N. Prakash, R. Venkitachalam, R. Cuthbert, A.R. Rahmani and A.A. Cunningham, 'Recent Changes in Population of Resident Gyps Vultures in India', *Journal of Bombay Natural History Society* 104, No. 2 (May–August 2007), 134.
15. Vibhu Prakash, interview with the author, 20 November 2013.
16. Ami Sedghi, 'Red List 2013: Threatened Species Across the Regions of the World', *The Guardian,* 26 November 2013, http://www.theguardian.com/news/datablog/2013/nov/26/iucn-red-list-threatened-species-by-country-statistics
17. International Union for Conservation of Nature (IUCN), 'IUCN Red List 2013', IUCN, https://docs.google.com/spreadsheet/ccc?key=0AonYZs4MzlZbdE9RY2hkX3lFYmNkQmFzTzI3Z0RXS0E&usp= sharing#gid=1, accessed on 6 March 2015.
18. Vibhu and Nikita Prakash, interviews with the author during site visit, 23–26 November 2009.
19. 'Report from the 3rd Meeting of Saving South Asia's Vulture's from Extinction (SAVE)', 2013, 81, http://www.save-vultures.org/save_latestnews.html#4thsavemtgreport

20. Sálim Ali, *The Illustrated Sálim Ali: The Fall of a Sparrow* (New Delhi: Oxford University Press, 2007), 8.
21. 'Drought and Downing Equal Vulture Supermarkets', Phys.org, 9 January 2014, http://phys.org/news/2014-01-drought-downing-equal-vulture-supermarkets.htmldoi:10.1371/journal.pone.0083470
22. Munir Virani and Patrick Benson, interviews with the author, 28 March 28, 2009
23. Munir Virani and Patrick Benson, interviews with the author, 29 November–2 December 2009.
24. Rameshwar, Pandevi, Ghanshyam Singh Naruka and Kashiram Bahkhar of the Ministry of Environment and Forests, and Jitu Solanki, interviews with the author, 3–5 December 2009.
25. World Health Organization (WHO) India, 'World Rabies Day 2013', WHO, http://www.searo.who.int/india/areas/communicable_diseases/rabies_day/en/, accessed on 6 March 2015.
26. Gardiner Harris, 'Where Streets Are Thronged with Strays Baring Fangs', *New York Times,* 6 August 2012, http://www.nytimes.com/2012/08/07/world/asia/india-stray-dogs-are-a-menace.html
27. Banjot Kaur Bhatia, 'No Mechanism to Dispose of Stillborn Babies in Patna Medical College', *Times of India,* 5 March 2014, http://timesofindia.indiatimes.com/city/patna/No-mechanism-to-dispose-of-stillborn-babies-in-Patna-Medical-College/articleshow/31435437.cms, accessed on 24 March 2015
28. Rachna Singh, 'EU Successfully Tracks Migration Path of Greater Steppe Eagle', *The Times of India,* 7 November 2014, http://timesofindia.indiatimes.com/city/jaipur/EU-successfully-tracks-migration-path-of-Greater-Steppe-Eagle/articleshow/45066192.cms
29. Saving Asia's Vultures from Extinction (SAVE), 'Conservation Story—The Problem', SAVE, http://www.save-vultures.org/save_conservationstory_background.html, accessed on 6 March 2015.
30. 'India's Ongoing War Against Rabies', Bulletin of the World Health Organization, 27, No. 12 (2009), http://www.who.int/bulletin/volumes/87/12/09-021209/en/, accessed on 24 March 2015.

31. Ibid.
32. Harris, 'Where Streets Are Thronged ...'
33. Tony Juniper, *What Has Nature Ever Done For Us?: How Money Really Does Grow on Trees* (London: Synergestic Press, 2013), Kindle book location 140–44.
34. Ibid.
35. M. Vijaikumar, D.M. Thappa and K. Karthikeyan, 'Cutaneous Anthrax: An Endemic Outbreak in South India', *Journal of Tropical Pediatrics* 48, No. 4 (2002), 225–26, http://www.ncbi.nlm.nih.gov/pubmed/12200984; Ramesh Reddy, Geetha Parasadini, Prasada Rao, Chengappa K. Uthappa and Manoj V. Murhekar, 'Outbreak of Cutaneous Anthrax in Musalimadugu Village, Chittoor District, Andhra Pradesh, India, July–August 2011', *Journal of Infection in Developing Countries* 6, No. 10 (2012), 695–699, http://www.jidc.org/index.php/journal/article/view/23103890; and 'State Announces Steps to Prevent Anthrax Outbreak', *The Times of India,* 9 January 2013, http://timesofindia.indiatimes.com/city/thiruvananthapuram/State-announces-steps-to-prevent-anthrax-outbreak/articleshow/17959464.cms
36. British Broadcasting Corporation (BBC), 'Case of Anthrax Confirmed in Lanarkshire Heroin User', BBC, 25 July 2012, http://www.bbc.com/news/uk-scotland-glasgow-west-18981196, accessed on 6 March 2015.
37. Yudhijit Bhattacharjee, 'Anthrax Has Hit Glasgow: The Story of a Desperate Hunt for Its Source', *WIRED,* 19 July 2012, http://www.wired.co.uk/magazine/archive/2012/08/features/anthrax-has-hit-glasgow
38. Mark Twain, *Following the Equator, A Journey Around the World,* Vol. 2 (New York: Harper & Bros, 1906), 53; and Joe Eaton, 'Silent Towers, Empty Skies', *Earth Island Journal,* Winter 2004, http://www.earthisland.org/journal/index.php/eij/article/silent_towers_empty_skies/
39. William E. Phipps, *Mark Twain's Religion* (Macon, Georgia: Mercer University Press, 2003), 198.
40. Ibid., 56.
41. Priyanka Pathak-Narain, 'Government to Offer Free Fertility

Treatments to Create Parsi Baby Boom', India Ink, *New York Times,* 18 December 2013, http://india.blogs.nytimes.com/2013/12/18/government-to-offer-free-fertility-treatments-to-create-parsi-baby-boom/?_php=true&_type=blogs&_r=2

42. In 2009 in Mumbai, 792 dead: 758 were consigned to towers and 34 were cremated, from photograph of undated *Parsiana* magazine seen by author at Dhan Baria's office.
43. Dhan Baria, phone interview with the author, 23 December 2006, and in-person interviews with the author, 10 December 2009 and 13 December 2013.
44. Ramola Talwar Badam, 'Photos of Funeral Ground Stir Anger', *Washington Post,* 7 September 2006, http://www.washingtonpost.com/wp-dyn/content/article/2006/09/07AR2006090700862_pf.html
45. Khojeste Mistree, interview with the author, 8 December 2009.
46. Khojeste Mistree, interview with the author, 13 December 2013.
47. Homi Dhalla, interview with the author, 16 December 2013.
48. Asad Rahmani, interviews with the author, 8 December 2009, and 17 December 2013.
49. Ananda Banerjee, 'It's Time to Look Beyond the Tiger', Live Mint, 14 January 2013, http://www.livemint.com/Politics/DSRW24PwlElXXUHjr978PM/Its-time-to-look-beyond-the-tiger.html
50. According to Rahmani and IUCN, http://www.iucnredlist.org/details/3129/0 and http://www.iucnredlist.org/details/22691932/0
51. Based on presentations, group discussions and interviews with the author, 7–9 November 2013.
52. 'Report from the 3rd Meeting of Saving Asia's Vultures from Extinction (SAVE)', 2013, 9.
53. Ibid., 59.
54. Vibhu Prakash, interview with the author, 20 November 2013.
55. Aditya Roy, 'Case Study: White Rumped Vulture (*Gyps bengalensis*)', Mohamed bin Zayed Species Conservation Fund, http://www.speciesconservation.org/case-studies-projects/white-rumped-vulture/6916, accessed on 6 March 2015.
56. 'Report from the 2nd Meeting of Saving Asia's Vultures from

Extinction (SAVE)', 2012, 12, http://www.save-vultures.org/save_latestnews.html#4thsavemtgreport

57. 'Report from the 3rd Meeting of Saving Asia's Vultures from Extinction (SAVE)', 54.
58. Ibid., 40–42.
59. Anil K. Sharma, Mohini Saini, Shambhu D. Singh, Vibhu Prakash, Asit Das, R. Bharathi Dasan, Shailey Pandey, Daulal Bohara, Toby H. Galligan, Rhys E. Green, Dietmar Knopp and Richard J. Cuthbert, 'Diclofenac is Toxic to the Steppe Eagle *Aquila nipalensis*: Widening the Diversity of Raptors Threatened by NSAID Misuse in South Asia', *Bird Conservation International* 24, No. 3 (2014), 282–86.
60. Richard J. Cuthbert, Mark A. Taggart, Vibhu Prakash, Soumya S. Chakraborty, Parag Deori, Toby Galligan, Mandar Kulkarni, Sachin Ranade, Mohini Saini, Anil Kumar Sharma, Rohan Shringarpure and Rhys E. Green, 'Avian Scavengers and the Threat from Veterinary Pharmaceuticals', *Philosophical Transactions of the Royal Society B,* 14 October 2013, http://dx.doi.org/10.1098/rstb.2013.0574
61. D. Swarup, R.C. Patra, V. Prakash, R. Cuthbert, D. Das, P. Avari, D.J. Pain, R.E. Green, A.K. Sharma, M. Saini, D. Das and M. Taggart, 'The Safety of Meloxicam to Critically Endangered *Gyps* Vultures and Other Scavenging Birds in India', *Animal Conservation* 10 (2007), 192–98, http://www.save-vultures.org/save_resources_scientificpapers.html
62. 'Report from the 3rd Meeting of SAVE', 34.
63. Ibid., 34–35.
64. Wildlife Extra, 'Vulture Killing Drug Now Available on EU Market', Wildlife Extra, March 2014, http://www.wildlifeextra.com/go/news/vulture-killing-drug-014.html#cr, accessed on 6 March 2015.
65. Based on presentations, group discussions and interviews with the author, 7–9 November 2013.
66. 'Report from the 3rd Meeting of SAVE', 80; and The Frozen Ark Project, http://www.frozenark.org/

5. AKASHA—ETHER

1. '*The Daily Show*—Extended Interview—Malala Yousafzai', YouTube video, https://www.youtube.com/watch?v=gjGL6YY6oMs, accessed on 6 March 2015.
2. Binod Singh and Pinki Kumari, interviews with the author, 5–6 November 2012.
3. United Nations Development Programme (UNDP), 'About Bihar', UNDP, http://www.in.undp.org/content/india/en/home/operations/about_undp/undp-in-bihar/about-bihar/, accessed on 6 March 2015.
4. Ibid.
5. Kounteya Sinha, 'Fertility Rate in India Drops by 19% in Ten Years', *The Times of India*, 1 April 2012, http://timesofindia.indiatimes.com/india/Fertility-rate-in-India-drops-by-19-in-10-yrs/articleshow/12487718.cms?referral=PM, accessed 24 March 2015.
6. Department of Economic and Social Affairs, 'World Population Prospects: The 2012 Revision database', United Nations, http://esa.un.org/wpp/unpp/panel_population.htm, accessed on 24 March 2015.
7. Sanjay Kumar, interview with the author, 7 November 2012.
8. World Health Organization (WHO), 'Maternal Mortality Country Profiles', Global Health Observatory (GHO) data, WHO, http://www.who.int/gho/maternal_health/countries/en/#I, accessed on 6 March 2015.
9. Office of the Registrar General, 'Maternal & Child Mortality and Total Fertility Rates—Sample Registration System', Office of the Registrar General, India, 7 July 2011, http://www.censusindia.gov.in/2011-common/Sample_Registration_System.html, accessed on 24 March 2015; and N.K. Singh and Nicholas Stern, *The New Bihar: Rekindling Governance and Development* (New Delhi: HarperCollins India, 2013), xxviii.
10. Madhuri Kumar, 'Over 69 Per cent Girls Married Before 18', *The Times of India*, 18 December 2012, http://timesofindia.indiatimes.com/city/patna/Over-69-girls-married-before-18/articleshow/17656917.cms, accessed on 24 March 2015.
11. Sara Sidner, 'India: The Pressure to Have Children', CNN,

22 April 2010, http://amanpour.blogs.cnn.com/2010/04/22/india-the-pressure-to-have-children/

12. Banjot Kaur Bhatia, 'Birth Rate per 1000 Population Is Highest in Bihar: Survey', *The Times of India,* 11 March 2014, http://timesofindia.indiatimes.com/city/patna/Birth-rate-per-1000-population-is-highest-in-Bihar-Survey/articleshow/31817135.cms
13. Nita Bhalla, 'India Advances, but Many Women Still Trapped in Dark Ages', Reuters, 13 June 2012, http://www.reuters.com/article/2012/06/13/us-g20-women-india-idUSBRE85C00A20120613
14. Firstpost India, 'Crude, Degrading Finger-Test Forced on Mumbai Gangrape Victim by Cops', Firstpost, 9 October 2013, http://www.firstpost.com/india/crude-degrading-finger-test-forced-on-mumbai-gangrape-victim-by-cops-1162157.html
15. Sandip Roy, 'The Long Journey: From Park Street Rape Victim to Suzette Jordan', Firstpost, 2 July 2013, http://www.firstpost.com/living/the-long-journey-from-park-street-rape-victim-to-suzette-jordan-921463.html
16. Swati Sengupta, 'For Kolkata Rape Victim, A Lonely Wait for Justice', India Ink blog, *New York Times,* 30 May 2013, http://india.blogs.nytimes.com/2013/05/30/for-kolkata-rape-victim-a-lonely-wait-for-justice
17. Gardiner Harris, 'For Rape Victims in India, Police Are Often Part of the Problem', *New York Times,* 21 January 2013, http://www.nytimes.com/2013/01/23/world/asia/for-rape-victims-in-india-police-are-often-part-of-the-problem.html?pagewanted=all&_r=0
18. Ellen Barry and Betwa Sharma, 'Many Doubt Death Sentences Will Stem India Sexual Attacks', *New York Times,* 13 September 2013, http://www.nytimes.com/2013/09/14/world/asia/4-sentenced-to-death-in-rape-case-that-riveted-india.html?ref=global-home
19. Beina Xu, 'Governance in India: Women's Rights', PBS News Hour online, 8 March 2013, http://www.pbs.org/newshour/rundown/2013/03/india-women.html
20. Human Rights Watch, 'India: Rape Victim's Death Demands

Action', Human Rights Watch, December 29, 2012, http://www.hrw.org/news/2012/12/29/india-rape-victim-s-death-demands-action, accessed on 15 March 2013.

21. Priya M. Menon, 'Lacking Support, Male Rape Victims Stay Silent', *The Times of India,* 16 February 2013, http://timesofindia.indiatimes.com/city/chennai/Lacking-support-male-rape-victims-stay-silent/articleshow/18524668.cms
22. Metropolitan Museum of Art, 'Mauryan Empire', Heilbrunn Timeline of Art History, Metropolitan Museum of Art, http://www.metmuseum.org/toah/hd/maur/hd_maur.htm, accessed on 6 March 2015.
23. Cheryl Colopy, *Dirty Sacred Rivers: Confronting South Asia's Water Crisis* (New York: Oxford University Press, 2012), 292.
24. Singh, N.K., and Nicholas Stern, (eds) *The New Bihar: Rekindling Governance and Development* (Noida: HarperCollins Publishers India, 2013), xx.
25. Indu Bharti, 'Bihar's Bane: Slow Progress on Land Reforms', *Economic and Political Weekly* 27, No. 13 (28 March 1992), 628–30.
26. Adolescents, interviews with the author, 5 November 2012.
27. Pathfinder, 'PRACHAR Phase I, II & III', Pathfinder, 26 October 2012.
28. United Nations Children's Fund (UNICEF), 'Child Marriage Is a Violation of Human Rights, but Is All Too Common', UNICEF, http://www.childinfo.org/marriage.html, accessed on 6 March 2015.
29. Anupam Srivastava and Jyoti Rao. 'Early Marriage: A Childhood Interrupted'. UNICEF, http://unicef.in/Story/1130/Early-Marriage—A-childhood-interrupted, accessed on 25 March 2015.
30. Nizamuddin Khan and Manas Ranjan Pradhan, 'Identifying Factors Associated with Maternal Deaths in Jharkhand, India: A Verbal Autopsy Study', *Journal of Health, Population, and Nutrition* 31, No. 2 (2013), 262–71, http://www.ncbi.nlm.nih.gov/pmc/articles/PMC3702348/
31. R. Mistry, O. Galal, and M. Lu, 'Women's Autonomy and Pregnancy Care in Rural India: A Contextual Analysis', *Social Science & Medicine* 69, No. 6 (2009), 926–33.

32. K.G. Santhya, Usha Ram, Rajib Acharya, Shireen J. Jejeebhoy, Faujdar Ram, and Abhishek Singh, 'Associations Between Early Marriage and Young Women's Marital and Reproductive Health Outcomes: Evidence from India', *International Perspectives on Sexual and Reproductive Health* 36, No. 3 (September 2010), http://www.guttmacher.org/pubs/journals/3613210.html
33. Rob Stephenson, Michael A. Koenig, Rajib Acharya and Tarun K. Roy, 'Domestic Violence, Contraceptive Use, and Unwanted Pregnancy in Rural India', *Studies in Family Planning* 39, No. 3 (September 2008), 177–86.
34. Mistry et al., 'Women's Autonomy and Pregnancy Care ...'; and Carla AbouZahr and Tessa Wardlaw, 'Maternal Mortality in 2000: Estimates Developed by WHO, UNICEF and UNFPA,' WHO, 2004.
35. Stephenson et al. 'Domestic Violence, Contraceptive Use ...'
36. Mara Hvistendahl, *Unnatural Selection: Choosing Boys Over Girls, and the Consequences of a World Full of Men* (New York: PublicAffairs, 2011), 87–88.
37. Praveen Swami, 'The Rapist in the Mirror', *The Hindu*, 22 January 2013, http://www.thehindu.com/opinion/lead/the-rapist-in-the-mirror/article4295240.ece
38. Office of the Registrar General and Census Commissioner India, 'Houselisting and Housing Census Data Highlights—2011', Census of India 2011, Government of India, 2012, http://www.censusindia.gov.in/2011census/hlo/hlo_highlights.html; and Dr C. Chandramouli, 'Housing, Household Amenities and Assets—Key Results from Census 2011', Census of India, Government of India, 2012, slide 74–76, http://www.censusindia.gov.in/2011census/hlo/Data_sheet/India/00_2011_Housing_India.ppt
39. Pavan K. Varma, *Being Indian: Inside the Real India* (New Delhi: Penguin Books, 2004), 36.
40. Vibhuti Patel, 'Tables on Sex Ratios in India and World', Academia, http://www.academia.edu/1741973/Tables_on_Sex_Ratios_in_India_and_World, accessed on 6 March 2015.
41. Office of the Registrar General and Census Commissioner India, 'Sex Ratios: Census 2001', Census 2011, Government of India,

http://www.census2011.co.in/census/state/haryana.html, accessed on 6 March 2015.

42. Valerie M. Hudson and Andrea M. den Boer, *Bare Branches: The Security Implications of Asia's Surplus Male Population* (Cambridge, MA: MIT Press, 2004), 125.
43. Bir Pal Singh, 'Sociology of Female Foeticide and Infanticide. Where Does the Law Stand?', *Gender Forum. An Internet Journal for Gender Studies* 38 (2012), http://www.genderforum.org/print/issues/passages-to-india/sociology-of-female-foeticide-and-infanticide/?fontsize=1?print=1
44. Elisabeth Bumiller, *May You Be the Mother of a Hundred Sons: A Journey Among the Women of India* (New York: Random House, 1990), 108.
45. Michelle Goldberg, *The Means of Reproduction: Sex, Power, and the Future of the World* (New York: Penguin, 2009), 174.
46. Hvistendahl, *Unnatural Selection,* 49.
47. Krishnan S. Nehra, 'Sex Selection & Abortion: India', US Library of Congress, http://www.loc.gov/law/help/sex-selection/india.php, accessed on 6 March 2015.
48. Prabhat Jha, Maya A. Kesler, Rajesh Kumar, Faujdar Ram, Usha Ram, Lukasz Aleksandrowicz, Diego G. Bassani, Shailaja Chandra and Jayant K. Banthia, 'Trends in Selective Abortions of Girls in India: Analysis of Nationally Representative Birth Histories from 1990 to 2005 and Census Data from 1991 to 2011', *Lancet* 377, No. 9781 (June 2011), 1921, http://www.sciencedirect.com/science/article/pii/S0140673611606491
49. Goldberg, *The Means of Reproduction,* 172.
50. 'Prenatal Sex Determination Clinics Flourishing in Delhi', *Outlook,* 3 July 2014, http://www.outlookindia.com/news/article/Prenatal-Sex-Determination-Clinics-Flourishing-in-Delhi/847986
51. International Center for Research on Women, 'Son Preference and Daughter Neglect in India: What Happens to Living Girls?', International Center for Research on Women, 2006, http://www.icrw.org/publications/son-preference-and-daughter-neglect-india
52. Hudson and den Boer, *Bare Branches,* 115.

53. International Center for Research on Women, 'Son Preference and Daughter Neglect in India'.
54. Sunny Hundal, *India Dishonoured: Behind a Nation's War on Women* (London: Guardian Books, 2013), 545–52.
55. Colin Freeman, 'Delhi Bus Rapist Blames His Victim in Prison Interview', *The Telegraph*, 1 March 2015, http://www.telegraph.co.uk/news/worldnews/asia/india/11443462/Delhi-bus-rapist-blames-his-victim-in-prison-interview.html, accessed on 25 March 2015.
56. Swami, 'The Rapist in the Mirror'.
57. Rafay Eajaz Hussain, interview with the author, 1 November 2012.
58. Sanjay Kumar, interview with the author, 7 November 2012.
59. Brigid Fitzgerald Reading, 'Education Leads to Lower Fertility and Increased Prosperity', Earth Policy Institute, 12 May 2011, http://www.earth-policy.org/data_highlights/2011/highlights13
60. Karthik Muralidharan and Nishith Prakash, 'Cycling to School: Increasing Secondary School Enrollment for Girls in India', International Growth Centre, 15 August 2014, http://www.theigc.org/project/cycling-to-school-increasing-high-school-enrollment-for-girls-in-bihar/
61. Pathfinder International, 'PRACHAR: Promoting Change in Reproductive Behavior in Bihar, India', Pathfinder International, http://www.pathfind.org/our-work/projects/prachar-promoting-change-in-reproductive-behavior-in-bihar-india.html, accessed on 6 March 2015.
62. Bumiller, *May You Be the Mother of a Hundred Sons,* 264.
63. Goldberg, *The Means of Reproduction,* 178.
64. Amartya Sen, 'India's Women: The Mixed Truth', *The New York Review of Books,* 10 October 2013, http://www.nybooks.com/articles/archives/2013/oct/10/indias-women-mixed-truth.
65. Sandeep Joshi, 'Child Marriage as Remedy for Rape Sparks Furore', *The Hindu,* 11 October 2012, http://www.thehindu.com/news/national/other-states/child-marriage-as-remedy-for-rape-sparks-furore/article3984623.ece
66. Goldberg, *The Means of Reproduction,* 183.

CONCLUSION

1. William Dalrymple, 'Narendra Modi: Man of the Masses', *New Statesman,* 12 May 2014, http://www.newstatesman.com/politics/2014/05/narendra-modi-man-masses?utm_source=
2. 'Narendra Modi: A Man of Some of the People', *The Economist,* 14 December 2013, http://www.economist.com/news/briefing/21591599-populist-nasty-past-and-decent-economic-record-wants-run-india-man-some
3. E.F. Schumacher, *Small Is Beautiful: Economics As If People Mattered* (New York: Harper & Row, 1973), 180.
4. Ananya Vajpeyi, *Righteous Republic: The Political Foundations of Modern India* (Cambridge, MA: Harvard University Press, 2012), xviii.
5. Arundhati Roy, *The Cost of Living* (London: Flamingo, 1999), 12.
6. Ellen Barry, 'After a Year of Outsize Expectations, Narendra Modi Ajdusts His Plan for India', *New York Times,* 25 May 2015, http://www.nytimes.com/2015/05/26/world/asia/after-a-year-of-outsize-expectations-modi-adjusts-his-political-course-for-india.html
7. Naazneen Karmali, 'Bill Gates in Bangalore', *Forbes,* 4 June 2012, http://www.forbes.com/sites/naazneenkarmali/2012/06/04/bill-gates-in-bangalore/
8. Ankit Singhi, 'Corporate Social Responsibility (CSR) Under New Company Law', Corporate Professionals, http://indiacp.blogspot.in/2013/01/CSR-Companies-Bill-2012.html, accessed on 6 March 2015.
9. Arpan Sheth and Anant Bhagwati, 'India Philanthropy Report 2013', Bain & Company, 5 March 2013, http://www.bain.com/publications/articles/india-philanthropy-report-2013.aspx

BIBLIOGRAPHY

Agrawal, G.D. 'An Engineer's Evaluation of Water Conservation Efforts of Tarun Bharat Sangh in 36 Villages of Alwar District', Alwar, Rajasthan: Tarun Bharat Sangh, 1996.

Ali, Sálim. *The Illustrated Sálim Ali: The Fall of a Sparrow*. New Delhi: Oxford University Press, 2007.

Bal, Hartosh Singh. *Waters Close Over Us: A Journey Along the Narmada*. Noida: Fourth Estate, 2013.

Banerjee, Abhijit V., and Esther Duflo. *Poor Economics: A Radical Rethinking of the Way to Fight Global Poverty*. New York: PublicAffairs, 2011.

Bang, Rani, Sunanda Khorgade, and Rupa Chinai. *Putting Women First: Women and Health in a Rural Community*. Kolkata: STREE, 2011.

Barnes, Douglas, Priti Kumar, and Keith Openshaw. *Cleaner Hearths, Better Homes: Improved Stoves for India and the Developing World*. New Delhi: Oxford University Press, 2012.

Berry, Wendell. *What Are People For?*. New York: North Point Press, 1990.

Boo, Katherine. *Behind the Beautiful Forevers: Life, Death, and Hope in a Mumbai Undercity*. New York: Random House, 2012.

Bumiller, Elisabeth. *May You Be the Mother of a Hundred Sons: A Journey Among the Women of India*. New York: Random House, 1990.

Carson, Rachel. *Silent Spring*. New York: First Mariner Books, 2002.

Chapple, Christopher Key, and Mary Evelyn Tucker, eds. *Hinduism and Ecology: The Intersection of Earth, Sky and Water*. New Delhi: Oxford University Press, 2000.

Chellaney, Brahma. *Water, Peace, and War: Confronting the Global Water Crisis.* New Delhi: Oxford University Press, 2014.

Choudhury, Chandrahas, ed. *India: A Traveller's Literary Companion.* New Delhi: HarperCollins India, 2010.

Colopy, Cheryl. *Dirty, Sacred Rivers: Confronting South Asia's Water Crisis.* New York: Oxford University Press, 2012.

Conway, Gordon. *One Billion Hungry: Can We Feed the World?.* Ithaca, NY: Cornell University Press, 2012.

Dharmadhikary, Shripad. *Mountains of Concrete: Dam Building in the Himalayas.* Berkeley, CA: International Rivers, 2008.

Deb, Siddhartha. *The Beautiful and the Damned: A Portrait of the New India.* New York: Faber and Faber Inc., 2011.

Dalmai, Vasudha, and Rashmi Sadana, eds. *Cambridge Companion to Modern Indian Culture.* Cambridge, UK: Cambridge University Press, 2012.

Doniger, Wendy. *The Hindus: An Alternative History.* New York: The Penguin Press, 2009.

Drèze, Jean, and Amartya Sen. *An Uncertain Glory: India and Its Contradictions.* Princeton, NJ: Princeton University Press: 2013.

Dubey, Krishna Gopal. *The Indian Cuisine.* New Delhi: PHI Learning Private Limited, 2011.

Eck, Diana L. *Banaras: City of Light.* Princeton, NJ: Princeton University Press, 1982.

Elliott, John. *Implosion: India's Tryst with Reality.* Noida, Uttar Pradesh: HarperCollins India, 2014.

Fagin, Dan. *Toms River: A Story of Science and Salvation.* New York: Bantam Books, 2013.

Flannery, Tim. *Here on Earth: A Natural History of the Planet Earth.* New York: Atlantic Monthly Press, 2010.

Fontanella-Khan, Amana. *Pink Sari Revolution: A Tale of Women and Power in India.* New York: W.W. Norton, 2013.

Friedman, Thomas L. *Hot, Flat, and Crowded: Why We Need a Green Revolution—And How It Can Renew America.* New York: Farrar, Straus and Giroux, 2008.

French, Patrick. *India: A Portrait.* New York: Vintage Departures, 2011.

Gadgil, Madhav, and Ramachandra Guha. *The Use and Abuse of Nature: Incorporating This Fissured Land: An Ecological History of India and Ecology and Equity.* New Delhi: Oxford University Press, 2000.

———. *Ecology and Equity.* London: Routledge, 1995.

George, Rose. *The Big Necessity: The Unmentionable World of Human Waste and Why It Matters.* New York: Metropolitan Books, 2008.

Giriharadas, Anand. *India Calling: An Intimate Portrait of a Nation's Remaking.* New York: Times Books, 2011.

Goldberg, Michelle. *The Means of Reproduction: Sex, Power, and the Future of the World.* New York: Penguin, 2009.

Harrison, Paul. *The Third Revolution: Population, Environment, and a Sustainable World.* London: Viking Penguin, 1993.

Hudson, Valerie M., and Andrea den Boer. *Bare Branches: The Security Implications of Asia's Surplus Male Population.* Cambridge, MA: The MIT Press, 2005.

Hundal, Sunny. *India Dishonoured: Behind a Nation's War on Women.* London: Guardian Books, 2013.

Hvistendahl, Mara. *Unnatural Selection: Choosing Boys Over Girls, and the Consequences of a World Full of Men.* New York: Public Affairs, 2011.

Jacob, Nitya. *Jalyatra: Exploring India's Traditional Water Management Systems.* New Delhi: Penguin, 2008.

Jacquet, Pierre, Rajendra K. Pachauri, and Laurence Tubiana, eds. *Development, the Environment and Food: Towards Agricultural Change?.* New Delhi: Agence Francaise de Developpement, the Institute for Sustainable Development and International Relations, and The Energy Resources Institute, 2011.

Jung, Anees. *Unveiling India: A Woman's Journey.* New Delhi: Penguin, 1987.

Juniper, Tony. *What Has Nature Ever Done for Us? How Money Really Does Grow on Trees.* London: Synergistic Press, 2013.

Kamdar, Mira. *Planet India.* New York: Scribner, 2007.

Kapur, Akash. *India Becoming: A Portrait of Life in Modern India.* New York: Riverhead Books, 2012.

Keown, Damien. *Buddhism, A Very Short Introduction.* Oxford: Oxford University Press, 2013.

Kolbert, Elizabeth. *The Sixth Extinction: An Unnatural History.* New York: Henry Holt, 2014.

Kristof, Nicholas D., and Sheryl WuDunn. *Half the Sky: Turning Oppression into Opportunity for Women Worldwide.* New York: Vintage Books, 2009.

Kumar, Amitav. *A Matter of Rats: A Short Biography of Patna.* New Delhi: Aleph Book Company, 2013.

Kumar, M. Dinesh. *Water Management in India: What Works, What Doesn't.* New Delhi: Gyan Publishing House, 2009.

Kumar, M. Dinesh, Nitin Bassi, A. Narayanamoorthy, M.V.K. Sivamohan, eds. *The Water, Energy and Food Security Nexus: Lessons from India for Development.* New York: Routledge, 2014.

Lawrence, Sir Walter Roper. *The Valley of Kashmir.* Henry Frowde, Oxford University Press Warehouse, London, 1895.

Leslie, Jacques. *Deep Water: The Epic Struggle over Dams, Displaced People, and the Environment.* New York: Picador, 2005.

Lopez, Barry, and Debra Gwartney, eds. *Home Ground: Language for an American Landscape.* San Antonio, Trinity University Press, 2006.

Luce, Edward. *In Spite of the Gods: The Strange Rise of Modern India.* New York: Anchor Books. 2006.

Mamdani, Mahmood. *The Myth of Population Control: Family, Caste, and Class in an Indian Village.* New York: Monthly Review Press, 1972.

Mander, Harsh. *Ash in the Belly: India's Unfinished Battle Against Hunger.* New Delhi: Penguin, 2012.

Miller, R. Eric and Murray E. Fowler. *Fowler's Zoo and Wild Animal Medicine: Current Therapy,* Vol. 7. St Louis, Missouri: Elsevier Saunders, 2012.

Mines, Diane P., and Sarah Lamb, eds. *Everyday Life in South Asia.* Bloomington: Indiana University Press, 2002.

Mishra, Anupam. *The Radiant Raindrops of Rajasthan.* New Delhi: Research Foundation for Science, Technology and Ecology, 2001.

Mistry, Cyrus. *Chronicle of a Corpse Bearer.* New Delhi: Aleph Book Company, 2012.

Nilekani, Nandan. *Imagining India: The Idea of a Renewed Nation.* New York: Penguin, 2008.

Nixon, Rob. *Slow Violence and the Environmentalism of the Poor.* Cambridge, MA: Harvard University Press, 2011.

Nordhaus, Ted, and Michael Shellenberger. *Break Through: From the Death of Environmentalism to the Politics of Possibility.* New York: Houghton Mifflin, 2007.

Pande, Mrinal. *Stepping Out: Life and Sexuality in Rural India.* Delhi: Penguin, 2003.

Radjou, Navi, Jaideep Prabhu, and Simone Ahuja. *Jugaad Innovation: Think Frugal, Be Flexible, Generate Breakthrough Growth.* San Francisco: Jossey-Bass, A Wiley Imprint, 2012.

Ringler, Claudia, Gerald C. Nelson, Christina Ingersoll, Ian Gray, Amanda Palazzo, Mark W. Rosegrant, Timothy B. Sulser, Tingju Zhu, Simla Tokgoz, and Richard Robertson. *Food Security, Farming, and Climate Change to 2050* [Kindle Edition]. International Food Policy Research Institute, 2011.

Ronald, Pamela C., and Raoul W. Adamchak. *Tomorrow's Table: Organic Farming, Genetics, and the Future of Food.* New York: Oxford University Press, 2008.

Roy, Arundhati. *The Cost of Living.* London: Flamingo, 1999.

Runge, C. Ford, Benjamin Senauer, Philip G. Pardey, and Mark W. Rosegrant. *Ending Hunger in Our Lifetime: Food Security and Globalization.*

Washington DC: International Food Policy Research Institute and John Hopkins University Press, 2003.

Sainath, P. *Everybody Loves a Good Drought: Stories from India's Poorest Districts.* New Delhi: Penguin, 1996.

Schumacher, E.F. *Small Is Beautiful: Economics As If People Mattered.* New York: Harper & Row, 1973.

Shiva, Vandana. *The Violence of the Green Revolution.* New York: Zed Books Ltd, 1997.

Shrivastava, Aseem, and Ashish Kothari. *Churning the Earth: The Making of Global India.* New Delhi: Penguin, 2012.

Singh, N.K., and Nicholas Stern, eds. *The New Bihar: Rekindling Governance and Development.* Noida: HarperCollins India, 2013.

Singh, Rajendra. *Rashtriya Jal Yatra (The Waterman's Journey).* Alwar, Rajasthan: Tarun Jal Vidyapeeth, 2005.

Swaminathan, M.S. *From Green to Evergreen Revolution: Indian Agriculture: Performance and Emerging Challenges.* New Delhi: Academic Foundation, 2010.

Turner, Chris. *The Leap: How to Survive and Thrive in a Sustainable Economy.* Toronto: Random House, 2011.

Vajpeyi, Ananya. *Righteous Republic: The Political Foundations of Modern India.* Cambridge, MA: Harvard University Press, 2012.

Varma, Pavan K. *Being Indian: Inside the Real India.* New Delhi: Penguin, 2004.

Vaidyanathan, A. *Water Resources of India.* New Delhi: Oxford University Press, 2013.

Weisman, Alan. *Countdown: Our Last, Best Hope for a Future on Earth?* New York: Little, Brown, 2013.

Williams, Florence. *Breasts: A Natural and Unnatural History.* New York: W.W. Norton, 2012.

Yadama, Guatam N. *Fires, Fuel & the Fate of 3 Billion: The State of the Energy Impoverished.* New York: Oxford University Press: 2013.

ABOUT THE AUTHOR

MEERA SUBRAMANIAN is a US-based freelance journalist whose work has been published in the *New York Times, Nature, Virginia Quarterly Review, Orion, Wall Street Journal* and other national and international publications. She earned an MA in journalism from New York University, and her writing on the disappearance of India's vultures received both the Staige D. Blackford Prize for Nonfiction and first place for outstanding feature story from the Society of Environmental Journalists Awards for Reporting on the Environment. She is also an editor of *Killing the Buddha,* an online literary magazine about religion, culture, and politics, and her essays have been anthologized in *Best American Science and Nature Writing* and *Believer, Beware: First-Person Dispatches from the Margins of Faith* as well as multiple editions of *The Best Women's Travel Writing.* Meera received a Fulbright-Nehru Senior Research Fellowship and a grant from the Fund for Environmental Journalism to support reporting for *Elemental India.* Find her at http://www.meerasub.org and @meeratweets. This is her first book.